DARK COMPANIONS OF STARS

Astrometric Commentary on the Lower End of the Main Sequence

PETER VAN DE KAMP

Reprinted from Space Science Reviews, Vol. 43, Nos. 3/4 (1986)

D. Reidel Publishing Company
Dordrecht / Boston

Library of Congress Cataloging-in-Publication Data

CIP data appear on separate card.

ISBN-13: 978-94-010-8586-1   e-ISBN-13: 978-94-009-4692-7
DOI: 10.1007/978-94-009-4692-7

Published by D. Reidel Publishing Company,
P.O. Box 17, 3300 AA Dordrecht, Holland.

Sold and distributed in the U.S.A. and Canada
by Kluwer Academic Publishers,
190 Old Derby Street, Hingham, MA 02043, U.S.A.

In all other countries, sold and distributed
by Kluwer Academic Publishers Group,
P.O. Box 322, 3300 AH Dordrecht, Holland.

To Ruth,
in memory of Olga and Maja

'Blessed are they that have not seen, and yet have believed.' *Saint John 20, 29.* (From Rembrandt's Gospel etchings.)

DARK COMPANIONS OF STARS

TABLE OF CONTENTS

PART II: UNSEEN STARS AND PLANETS

PREFACE

If you want to understand the invisible, look careful at the visible.

The Talmud

A 'bird's eye' or rather a distant spacecraft's view of the solar system reveals an assembly of planets, terrestrial, giant and Pluto. The orbital motions are in the same sense, counter clockwise, as seen from the north of the general flattened space within which the planetary motions are confined. This state of affairs is corevolving and, more or less, coplanar. The rotations are in the same sense as the revolutions, with the striking exception of Uranus whose sense of rotation is perpendicular to its plane of revolution. As time goes by, most of the planets remain fairly close to a general plane and at no time stray unduly far from it; they remain confined within a rather narrow box or *disk* with a large 'equatorial' extent. The most distant planet, Pluto, requires a diameter of some 80 astronomical units for the disk. One astronomical unit is the distance of the Earth to the Sun, to be more precise the length of half the major axis of the Earth's slightly elliptical orbit around the Sun, and amounts to nearly 149 600 000 km.

For the comets in our solar system the situation is different. Their orbits are not organized in any systematic manner; they are neither co-revolving nor co-planar. With the passage of time the comets remain within a huge cloud surrounding the Sun, roughly 100 000 astronomical units or over half a lightyear in diameter. The 'arrangement' and 'state of motion' of planets and comets find their galactic equivalents, on a very grand scale, in the 'populations' of our Milky Way System; we may speak of a *halo* of comets, as contrasted with the 'disk' of terrestrial and giant planets.

This treatise consists of two main parts. Part I surveys our *cosmic neighborhood*, the 'fixed stars' and the 'wandering' planets (Chapter 1), the motions and distances of stars (Chapters 2 and 3), the physical properties of stars (Chapter 4). The nearby stars are discussed as in their striking division into red and white dwarfs (Chapter 5). Of particular interest for the purpose of this article is a more detailed description of the intrinsically faintest stars, both visible and invisible (Chapter 6). The *photographic astrometric* survey of the lower end of the Main Sequence leads to information about unseen companions of stars, particularly the *dark dwarfs*. Next follows a survey of the solar system, more specially the planetary system of our Sun. Its geometric, kinematic, physical and chemical properties are presented (Chapter 7), followed by its possible origin, as well as that of stars and further cosmic assemblies. An attempt is made to define stars and planets (Chapter 8).

Part II deals with *unseen stars and planets*. Early historical discoveries include the perturbations of Sirius and Procyon, Neptune, a section on the discovery of Pluto and a search for a further planet X (Chapters 9 and 10). A description of the recent and current photographic astrometric approach follows. Companions from perturbations, technique, analysis and interpretation are discussed (Chapters 11–14). The status of nearby Barnard's star is considered in Chapter 15. These studies are a technical sequel to the determination of stellar parallaxes (Chapter 3) luminosities and masses (Chapter 4). All these are precision aspects of astrometry in the photographic or visual radiation, the space-time properties of stellar paths. More than the past half century has witnessed the continuing developments; increased telescopic and photographic power and improved measuring apparatus. A relatively recent technique, a valuable addition to the gravitational approach, is the *infrared speckle interferometry*.

For further astrometric information the reader is referred to *Principles of Astrometry* (W. H. Freeman & Co., 1967) and to *Stellar Paths* (D. Reidel Publishing Company, 1981).

I am indebted to Robert S. Harrington, John L. Hershey, Wilhelm Gliese, and Hans-Jürgen Treder for their continued and inspiring helpfulness. Particular appreciation goes to Clyde M. Tombaugh for the personal account of his discovery of Pluto (Chapter 10).

INTRODUCTION

The most basic astronomical observations are those giving the positions and motions of stars.

Because of the motion of the Earth in its orbit around the Sun, the vantage point from which we observe the stars changes; the position of a nearby star changes when compared with positions of more distant stars. Determination of the parallax of the nearby stars is the fundamental method of determining stellar distances. From these distances come calibrated intrinsic brightnesses of many kinds of stars and, in careful extrapolations, ultimately the distance scale and age of the Universe.

Because stars have motions of their own through space, they change position with respect to each other on the celestial sphere. The angular rate of motion of a star across the line-of-sight is its proper motion. From the proper motion of a star and an estimate of its distance follows one component of the intrinsic velocity of the star through space. Measurements of proper motions are essential to study the structure and dynamics of our Galaxy. Determination of the galactic force field implied by stellar motions remains a problem of particular importance.

Because many, probably most, stars occur in pairs or multiples orbiting a common center of gravity, their paths through space may reveal a wobble. Usually only a single image of the system occurs on a photographic plate, but if followed long enough perturbations to its proper motion may reveal the orbits of the members of the system. All fundamental determinations of stellar masses are provided by application of Kepler's law to the orbits of binary systems; we have no direct way of knowing the mass of an isolated star. Stellar evolution and structure are essentially determined by stellar mass.

Prof. Peter van de Kamp has given us here a review which shows why the craft of astrometry, of which he is the master, requires that its practitioners be meticulous, patient, and persistent. The apparent motions of stars are painfully slow: although Hipparchus was able to measure the parallax of the Moon and thus its distance from the Earth, it was two thousand years later, in 1838, that Friedrich Wilhelm Bessel determined the first stellar parallax, for the star 61 Cygni. Proper motions were not discovered until 1718, when Edmund Halley noticed the displacement of Sirius and Arcturus from the positions recorded by Ptolemy in the second century AD. Perturbations to the proper motions of Sirius and Procyon were noted in 1844 by Bessel and attributed to the presence of unseen companions which later observations revealed.

Most of the modern parallax and proper motion results come from the techniques of long-focus photographic astrometry, long championed by Peter van de Kamp at the Sproul Observatory of Swarthmore College and described in this book. With current techniques parallaxes are routinely measured to distances of 50 pc; at greater distances uncertainties are large. Thousands of stars have annual proper motions exceeding 0.5 sec of arc; accuracies as high as 0".001 annually are available for a few stars, observed over decades. The slow motions of stars and the necessity of high accuracy require that research programs be diligently pursued over many years and that their integrity be carefully guarded. Of course, some of the astronomical goals will be extended by other techniques. The Hipparcos satellite under development for launch in 1988 by the European Space Agency will bring within range of accurate parallax measurements during its two years of operation many types of stars which are too rare to be represented in our immediate neighborhood. Techniques of radio interferometry now give positions of radio sources with accuracies similar to that available for stars; measurement of the proper motion of the nucleus of our Galaxy will give a direct measure of the galactic rotation speed near the Sun. Astrometric analyses are also being extended to such exotic objects as binary pulsars and suspected black holes. Measurements of the orbital decay of the binary pulsar offer confirmation of Einstein's general theory of relativity. The short time-scale of variation in the Cygnus X-1 source indicates a binary system involving an object so compact that it is plausibly a black hole. But for so much work there is no substitute for prolonged, uninterrupted ground-based observations.

Continuity of effort is especially crucial for the searches, so thoroughly described in this work, for unseen stellar companions. The orbital periods of known double stars are typically tens or hundreds of years. Accurate measurements made over a few years might reveal perturbations from uniform rectilinear motion, but more time must pass before reliable orbital elements can be determined. It is just this work, too, that has the special importance of giving not only the masses of stars but also the discovery and masses of unseen companion objects which may not be stars at all. On both observational and theoretical grounds astronomers believe that there is far more mass in the Universe than is visible. Whether or not Jupiter-like objects, of mass insufficient to ignite as a star, or the brown-dwarf cinders of burnt out stars, contribute in any important way to the missing mass are questions which motivate the continued astrometric search for unseen stellar companions. The search for true planetary systems has a special attraction of its own, of course, and is marvelously told here for the case of Barnard's star.

There is another aspect of continuity which is of great importance to Peter van de Kamp. He recognizes the continuity of scholarly effort and is thrilled to be part of it. He respects deeply the work of Frank Schlesinger and others establishing the techniques of long-focus photographic astrometry, and is fully aware that studies of stellar motions have been crucial to what he calls the galactocentric revolution. Astrometric work led to the discovery by Jacobus Kapteyn early in this century of the preferential motion of stars in the directions of Sagittarius and Auriga. By 1920 Harlow Shapley had recognized the nature of the Kapteyn stellar system as a spiral galaxy, and had estimated its dimensions. The theory of galactic rotation with the feature of subsystems rotating at different speeds was developed by Bertil Lindblad in 1926. In 1927 Jan Oort discovered differential galactic rotation and gave it a mathematical formulation which successfully explained the preferential motions as well as the asymmetries in the distribution of stars of high velocity. In all this work accurate astrometric work was indispensable.

Peter van de Kamp reveres the continuity of the astronomical scholarly tradition. His inventory of the nearby stars and their motions stands as a monument, but to the many of us who cherish his example this reverence is as much a part of what he gives us.

September 1985   W. B. BURTON
Leiden Observatory

CHAPTER 1

THE 'FIXED' STARS AND THE 'WANDERING' PLANETS

A. Celestial Sphere. Constellations

To the unaided eye the starry sky reveals a few thousand stars 'attached' to the illusion of a celestial sphere, the night sky. The positions i.e. directions in which we observe stars, planets and other objects are referred to a fundamental great circle and a zero point on that circle. In the basic *equatorial* coordinate system these are the celestial equator, an extension of the Earth's equator, and the vernal equinox at which the Sun in its apparent annual orbit ascends from south to north of the celestial equator. The two equatorial coordinates of an object are *Declination* (Decl.), measured along the great circle, named *hour-circle* through the object and the celestial poles (extensions of the Earth's axis of rotation); *Right Ascension* (RA) is measured from the vernal equinox eastward to the intersection of hour-circle and equator. Both coordinates are referred to as *fundamental* or *'absolute'* positions.

As a result from the gravitational action of Moon and Sun on the equatorial bulge of the Earth, the Earth's axis of rotation describes a conical motion with a cycle of 25.800 years. Hence the equinox shifts 50 arc sec yearly. This phenomenon named *precession* was discovered by Hipparchus as early as 125 BC, it leads to continually changing equatorial coordinates of celestial objects. Note that these changes are due to the gradually changing reference system, not to the observed objects. True shifts in stellar positions, their angular *proper motions* were not discovered till 1718. Up to that time the stars were truly considered 'fixed', more about this in Chapter 2. Even now the proper motions of the great majority of stars are so small that to the unaided eye no apparent changes in the constellations are detectable during a man's life time. To study the proper motions of stars, precession is allowed for. Man's sense of design, aided by his imagination, has grouped the brighter stars together into *constellations*, and these appear, at first sight, to remain undisturbed not only within themselves but in relation to one another. Though the stars in constellations may seem to 'belong' to each other, any such connection generally is fortuitous. The grouping together is based on more or less common direction and brightness only, and does not *a priori* imply any physical relation between the stars, which may be at entirely different distances from us. It appears however, as if the stars, all stars, and hence the constellations are attached or fixed to the celestial sphere, a persistent illusion, very convenient for descriptive purposes.

Though the constellations, appear to be fixed on the *celestial* sphere truly, they are not fixed and both the distant future as well as the distant past present differently shaped constellations because of the proper motions of stars. To the unaided eye, there is no noticeable change even over centuries, and the 'fixed' stars remain a suitable background on which the motions of the 'wandering' planets may be studied as observed from our planet 'Earth'. It is all a matter of distance in this 'restless Universe', the nearest star being over a hundred thousand times further away than say a planet like Mars. No wonder that the slowly changing stellar positions were not discovered until the beginning of the eighteenth century, while the changing positions of the planets with respect to the starry background had been always obvious, being easily detectable with the unaided eye, often from night to night, for comparatively nearby *planets* like Venus and Mars.

An interesting question thus arises in connection with the stellar constellations, which assume different shapes over long time intervals of thousands or tens of thousands years in the past or in the future. The present author and presumably others too, are firmly of the opinion that the current forms ('gestalt') of the constellations are 'perfect' – just as they 'should be'. Look for example at the Big Dipper (Ursa Major); its shape is beautiful, the locations and spacings of the seven bright stars are 'just right'. Is this a question of 'growing accustomed to its face', or 'man gewöhnt sich d'ran', or an instinctive loyalty to things as they are during our brief sojourn on the Earth? During one life time, i.e. up to an interval of one century, there is no detectable change in the constellations as they appear to the naked eye. But over intervals covering tens of thousands of years or more in the future or in the past the situation changes markedly, and the constellations appear distorted from what they appear now.

Is this another example of man's desire, instinct or what have you, to be the center of the Universe, in space as well as in time? Is it not logical to assume that early man, or later man, separated from us by tens or hundreds thousands of years, is equally pleased with his constellations and will frown and pity us for, from his point of view, distorted constellations, which right now to us appear just right and perfect?

After all, this is an excellent example of *relative* if not homocentric point of view, there being no preferred location in space and time, only an unavoidably, inexorably determined viewpoint from any limited location and epoch, at which the observer is bound to live, and enjoy.

B. Copernican Revolution

Long ago Aristarchus of Samos, ± 310–230 BC, considered the idea of the Sun occupying a central position among the planets. The *geocentric* viewpoint of Ptolemy persisted however: the Earth remained the center of description of planetary motions. While the general direction of these motions on the sky is west-to-east, part of the time the planets appear to move from east to west. The motion of each planet was interpreted by Ptolemy as the resultant of yearly circular motion (epicycle) around a point on a large circular orbit (deferent) with the Earth at its center.

No satisfactory picture of the 'true' state of planetary motions was obtained until the middle of the sixteenth century. The essential difficulty was that the observer on the Earth is part of the cosmic experiment which he is observing; at all times he is travelling along this path i.e. orbital motion around the Sun. Moreover the planetary assembly, including the Earth, does not permit a bird's eye view of the situation; all motions take place close to the same plane, thus hindering the analysis.

It required the imagination, persistance and courage of Nicolaus Copernicus (1473–1543) to overcome these difficulties; in 1543 he demonstrated that the state of motion of the planets, including the Earth, can be described very simply if the observer imagines himself to be viewing the planets from the Sun. In other words, if in our imagination we transfer the observer to the location of the Sun, the apparent complexity of planetary motions, as seen from the Earth disappears entirely. Sun and planets follow a single simple pattern of motion, the orbits are virtually circular and differ in size. The general sense of direction of travel is the same for all planets, counter clockwise as seen from north of the orbital planes which differ little from each other; in other words the planetary system, at least the terrestrial and the giant planets, at all time essentially can be fitted into a flattened space pocket or *disk*.

As observed from the Earth, our view of the planets shifts throughout the year. The predominant eastward or *direct* motion alternates with a part-time westward or *retrograde* motion. The retrograde motions of the planets beyond the Earth's orbit are centered around the time of their closest approach to the Earth; the duration of these retrograde motions ranges from less than three months for the nearest outer planet Mars, to almost half a year for the most distant planet Pluto. Closest approach occurs near *opposition*, i.e. the time that planet and Sun are in nearly opposite directions as seen from the Earth.

The observed angular sweep of the retrograde motions decreases for planets at greater distances from the Sun. These reflex motions are a beautiful revelation of the Earth's orbital motion and of the general spatial arrangement of the planetary orbits. It was not until the middle of the nineteenth century that the equivalence of these planetary reflex motions was observed for the much more distant *stars*. This, a routine procedure, nowadays furnishes the basic approach to determining stellar distances (Chapter 3). The situation for the inner planets Mercury and Venus, is more complicated because of the rapid orbital motions of these objects.

C. Kepler's Laws. Newton's Law of Gravitation

While the heliocentric revolution had thus irrevocably taken place and established itself, further adjustments were indicated. Copernicus adopted circular orbits, more or less in the same plane, but their centers were not at the precise location of the Sun. Moreover the orbits showed slight deviations from being circular. The worship of the 'perfect' circle had to abandoned a little more than half a century later, through the efforts and analysis by Johannes Kepler (1571–1630), who established his famous three laws. His first two laws (1609) describe the motions of the planets as taking place in *elliptical* orbits with the Sun in focus subject to the *law of areas*, which states that within any one planetary orbit the line (radius vector) joining Sun and planet, sweeps over equal areas of space in equal intervals of time. *The law of ellipses* and the *law of areas* were followed by Kepler's *third or harmonic law* (1619), the simple relation for all planets that the cube of the mean distance from the Sun is proportional to the square of the duration or period of one orbital revolution.

Kepler's laws resulted from an analysis of long series of observations by Tycho Brahe (1546–1601) of the changing locations of the planets on the starry sky during the second half of the sixteenth century. The optical telescope was not invented till 1608. It is a tribute to the human mind that such significant deductions could be and were made. Kepler's laws rank among the greatest scientific discoveries of all time.

The breakthroughs of Copernicus and Kepler led to the formulation of the universal *law of gravitation* by Isaak Newton (1642–1727) toward the end of the seventeenth century. The Sun is considered the 'cause' of the accelerated motion, the continuous 'falling', of planets and comets, toward or rather *around* the Sun. This *gravitation* is also exhibited 'locally' by the falling around the Earth of the Moon or any artificial satellite, or of falling objects near the surface of the Earth, referred to as gravity. The rate of falling depends on the distance of the object from the center of the Earth which obviously 'causes' the falling. Meanwhile our knowledge about motions of terrestrial objects had been crystallized in the so-called *laws of motion* of Newton (though others had contributed also). In slightly revised version these may be stated as the following postulates:

(1) An object sufficiently far away from another object moves in a straight line with uniform speed i.e. suffers no change of motion or acceleration (also named *Law of Inertia*).

(2) The acceleration of any object is caused by the presence and influence of another sufficiently nearby object. This acceleration may be attributed to a *force* defined as *mass times acceleration*. This concept of force based on terrestrial human experience does not explain but simply describes gravitation.

(3) Any two objects influencing each other do so with equal and opposite force, which means that the accelerations which they experience are inversely proportional to their masses, and in opposite directions. This is the law of *action equalling reaction*.

Postulate 1 implies constant areal velocity with respect to any point, and thus is the equivalent of Kepler's law of equal areas. Conversely Kepler's law of equal areas may be referred to as *elliptical inertia* contrasted with *linear inertia* for single objects. (*Astronomy and Astromechanics, Popular Astronomy*, Vol. 52, No. 2, 30–46, February 1944.)

A simple derivation of Newton's law of gravitation follows. Through mathematical analysis the first two laws of Kepler lead to the simple conclusion that for any one planetary orbit the attraction, i.e. central acceleration from planet to Sun, varies with the inverse square of the distance r between Sun and planet. The third law leads to the conclusion that this central acceleration f varies with the inverse of the distance r between Sun and *any* planet. This last deduction is easily derived for the simplified case of circular orbits, in which case the well known kinematical relation exists between central acceleration for orbital velocity V and radius r of the orbit

$$f = \frac{V^2}{r} \tag{1.1}$$

(the units are c.g.s.).

Kepler's third law may be written as

$$\frac{r^3}{P^2} = \text{constant} \tag{1.2}$$

where P is the period of revolution. Or, since

$$P = \frac{2\pi r}{V} ,$$

we deduce

$$V^2 (:) \frac{1}{r} . \tag{1.3}$$

Formula (1.3) expresses the simple fact that for the idealized case of circular motion the orbital velocities of the planets vary with the inverse square root of their distance from the Sun.

Combining the kinematical relation (1.1) with the observed cosmic relation (1.3) we see that

$$f(:) \frac{1}{r^2} . \tag{1.4}$$

This is the famous inverse square law for the 'diluting' of the falling attraction with increasing distance from the Sun.

The Sun therefore is the 'cause' – the source of attraction for the planets. Conversely, and this is one of Newton's laws of motion: *action equals reaction*, the inverse attraction must be mutual, i.e., the planets are also a source of attraction which results in a minute 'falling' of the Sun toward them. The inverse square acceleration is therefore assumed to be mutual; that is, not only does the planet 'fall' toward the Sun, but the Sun also 'falls' toward the planet, or rather, both fall, from opposite sides, toward their equilibrium point, the center of mass. The observed *relative* or mutual acceleration f_r is, in absolute value, the sum of the acceleration f_m of the planet and of the much smaller acceleration f_M of the Sun; that is

$$f_r = f_m + f_M . \tag{1.5}$$

Generalizing the relation for any two particles with masses $\mathcal{M}_1$ and $\mathcal{M}_2$, each of the two particles may be considered the center of a *field* of gravitational attraction of spherical symmetry. At any location the amount of the falling acceleration toward the center depends on the central mass, and on the distance from (the center of) that mass. What happens to the masses $\mathcal{M}_1$ and $\mathcal{M}_2$, separated by distance r? Along this segment r:

$\mathcal{M}_1$ experiences an acceleration $f_1 = G \dfrac{\mathcal{M}_2}{r^2}$ toward $\mathcal{M}_2$,

$$\tag{1.6}$$

$\mathcal{M}_2$ experiences an acceleration $f_2 = G \dfrac{\mathcal{M}_1}{r^2}$ toward $\mathcal{M}_1$.

Hence the relative or mutual attraction i.e. the total effective falling acceleration between $\mathcal{M}_1$ and $\mathcal{M}_2$ is the sum of two respective falling rates i.e.

$$f = f_1 + f_2 = G\,\frac{\mathcal{M}_1 + \mathcal{M}_2}{r^2}\,. \tag{1.7}$$

This is Newton's law of universal gravitation, where $G = 6.67 \times 10^{-8}$ (c.g.s. units) is Newton's *constant of gravitation*.

While derived from planetary motions, Newton's law is generalized for particles and for objects possessing spherical symmetry, a frequent occurrence to a high degree of approximation for celestial bodies.

Introducing the concept of force, F we have for the equal forces of action and reaction

$$\mathcal{M}_1 f_1 = \mathcal{M}_2 f_2 \tag{1.8}$$

which leads to the expression

$$F = G\,\frac{\mathcal{M}_1\mathcal{M}_2}{r^2}\,. \tag{1.9}$$

However since force is not primarily an observable concept in the study of cosmic motions, but acceleration *is*, expression (1.9) is particularly significant and useful in astronomical problems.

CHAPTER 2

STELLAR MOTIONS

A. Proper Motions

It was not until 1718 that the illusion of 'fixed' stars was destroyed. In that year Edmund Halley (1656–1752) announced that the bright stars Sirius, Betelgeuse, Aldebaran and Arcturus had moved appreciably from the positions recorded by Ptolemy in the second century A.D. The rates of these *proper motions* are known with high accuracy. For Betelgeuse and Aldebaran however the annual proper motions are now known to be very small, $0\overset{''}{.}03$ and $0\overset{''}{.}20$ respectively and Halley's results for these stars must be considered spurious. The annual displacements of Sirius and Arcturus are quite large, amounting to $1\overset{''}{.}32$ for the former and $2\overset{''}{.}29$ for the latter. The accumulated displacements over the fifteen centuries between Ptolemy and Halley are therefore 33' for Sirius, and as much as 57' for Arcturus, both exceeding the apparent diameter of the Moon. Since Newton's law of gravitation was established, it should be no surprise – on the contrary, it had to be expected that stars must be in some state of motion, to counteract a possible collapse due to gravitation.

Gradually more and more stars were found to be moving. At present proper motions are known for hundreds of thousand stars. The rate of motion of a single star remains the same as time goes by. Exceptions of specific interest are to be discussed later. While the majority of stars prove to have small motions, large proper motions, presumably of nearby stars, are sufficiently frequent to give the stellar world an aspect of restlessness. The number of known stars of large proper motion includes a large proportion of apparently faint stars. For known visual binaries, this constancy holds for the center of mass of the components, while deviations from the constancy of apparently single stars point to unseen companions (Chapter 11).

The cumulative effect of time plays an important role in the accurate determination of stellar motions. Telescopic observations of the locations of thousands of stars have been carried on for more than three centuries. These long-range studies are made with meridian or transit circles, telescopic instruments equipped with vertical, finely graduated dials. The instrument is mounted in such a way that the transit of a star may be observed when the diurnal rotation of the celestial sphere carries the star past the meridian. The altitude of the star can be accurately observed and the time of meridian transit noted on a sidereal clock. Observations of this type permit an accurate mapping of the stars on the celestial sphere down to about the tenth magnitude. From a comparison of the right ascensions and declinations obtained at different times, the proper motions of the stars may be then be derived. The photographic approach has also proven useful in conjunction with visual observations.

The positions and proper motions of fainter stars may be derived photographically by comparing photographs taken at different epochs. Through the efforts of Max Wolf

(1863–1932), Frank E. Ross (1874–1960), Willem J. Luyten (1899–   ), Henry
L. Giclas (1910–   ) and others, thousands of stars of 'large' proper motion have thus
been found. These objects with annual proper motions, say larger than $0\rlap{.}''2$ are detected
and measured on the background of abundantly available, presumably distant stars of
comparatively negligible motion.

B. Space Motions

The angular proper motion of a star may be converted to linear measure if the parallax
of the star is known. The problem of parallax determinations is discussed in the next
chapter. Since the angular parallax value p represents the linear value of one unforeshortened astronomical unit, the annual proper μ corresponds to μ/p astronomical units
per year. Or since

$$1 \text{ AU} = 149.6 \times 10^6 \text{ km}$$

and one year is 3.156×10^7 seconds of time, we have the equivalence

$$1 \text{ AU yr}^{-1} = 4.74 \text{ km s}^{-1} . \tag{2.1}$$

Hence the linear cross-section or *transverse velocity* component. T, perpendicular to the
line-of-sight, amounts to

$$T = 4.74 \, \frac{\mu}{p} \text{ km s}^{-1} . \tag{2.2}$$

Some find it convenient to derive and remember relation (2.2) as follows. The Earth's
orbit is nearly circular; the Earth with its orbital velocity of 29.8 km s^{-1} around the Sun,
travels 2π astronomical units yearly. We thus have, again

$$1 \text{ AU yr}^{-1} = \frac{29.8}{2\pi} = 4.74 \text{ km s}^{-1} . \tag{2.3}$$

An example: for Barnard's star

$$\mu = 10\rlap{.}''31 , \qquad p = 0\rlap{.}''547 ,$$

hence

$$T = \frac{\mu}{p} = 18.1 \text{ AU yr}^{-1} \quad \text{or} \quad T = 4.74 \times \frac{10.31}{0.547} = 89 \text{ km s}^{-1} . \tag{2.4}$$

Motion perpendicular to the sky, i.e. in the line-of-sight, the *radial velocity* is measured
spectroscpically with the aid of the Doppler effect (Christian Johann Doppler,
1803–1853).

When both transverse T and radial motion V are known, the space motion

$$v = \sqrt{T^2 + V^2}$$

may be obtained. Spectroscopic observations establish for example, that Barnard's star has a motion of approach of 108 km s^{-1}. The resultant of the cross motion and the radial velocity, computed as the hypothenuse of a right triangle is

$$\sqrt{89^2 + 108^2} = 140 \text{ km s}^{-1} \quad \text{or} \quad 40 \text{ AU yr}^{-1}$$

a comparatively high space velocity for a star. Barnard's star is one of the few stars for which a slow, gradual perspective change in the proper motion can be observed. The proper motion increases each year by 0''.0013; in about ten thousand years the star will be nearest to us. At that time its distance will be slightly under four lightyears, and it will appear one magnitude brighter than it is now (Chapter 15).

A great deal of interesting information on stellar motions has been acquired. We find stars that move slowly, others that move fast, the average speed being something like 30 km s^{-1}. We find, too, that different stars move in different directions (Section D).

C. Law of Inertia

In studying proper motion we ignore or allow for the stellar annual parallactic effect due to the Earth's orbital motion around the Sun. The context of our observations therefore involves the motion of the star with respect to the Sun. There is no 'real' rest of the star or any other point. We have developed convenient conventional notions of rest. The rotation and revolution of the Earth have already given us perspective on the relativity of rest and motion. All we can observe is relative locations, relative paths of different bodies with respect to each other; the relativity of these observations alone is 'real'. Any concept of rest is simply an arbitrary notion. The heliocentric revolution has transferred our viewpont from the Earth to the Sun. This new vantage point is chosen for convenience and simplification in grasping the mechanics underlying planetary and stellar motions. One initial result of the study of stellar motions is outstanding in simplicity and importance. The space motions of the vast majority of stars as observed from the Sun take place in straight lines and with uniform speed. The validity of this statement depends, of course, on the margin of accuracy and space-time limitations of our observations. However, it appears, at first sight at least, that stars obey the simplest form of motion imaginable. This law of stellar motions implies, therefore, the virtual absence of any observable acceleration between Sun and stars. This property of stellar motions is the equivalence of the law of inertia first inferred from terrestrial experiments. This law, already recognized by Galileo and Newton, may be expressed as follows: a body which is sufficiently far away from other bodies undergoes no acceleration.

This important universal law of motion is beautifully exhibited by the stars. Its illustration in terrestrial experiments is always complicated by the disturbing effects of friction and of gravity, i.e., by the Earth. A smooth billiard ball rolling on a smooth horizontal table provides about a good terrestrial experiment for demonstrating the law of inertia. However, proper support against gravity must be provided, while friction can be only partly eliminated! In the stellar world the qualifying condition – sufficiently far away from other bodies – generally is sufficiently satisfied to permit the operation of the law of inertia for a free body, unsupported and apparently not subject to any measurable friction. The cosmic equivalence of these billiard balls are the stars rolling along in the great open spaces of the Universe, each with a uniform speed in a constant direction. At least at first sight.

Qualifying conditions for the law of inertia, namely are a reasonably limited interval of time and an appropriate choice of background. As a rule these conditions are satisfied in astronomical observations and we need not be concerned about them any further at this time.

Our view of the Universe is handicapped by our location. We live at the bottom of our atmosphere, which dims the light messages from the cosmos; moreover, any motion in our immediate experience is strongly affected by the gravitational prison which rather confines us to the surface of the Earth. Biologically, both atmosphere and gravity are necessarities; but they do frequently complicate and interfere with our understanding of the inanimate world of astronomy. 'Space' research may remove some of the difficulties, except for the time factor which is independent of any technical approach.

While the law of inertia appears to be beautifully demonstrated by Sun and stars, and (Chapter 11) by the mass-centers of binary stars, exceptions exist. First recognized in 1844 (Chapter 9) these exceptions by no means contradict the law of inertia. On the contrary the apparent deviations draw attention to and are simpliy explained by the qualifying condition in the law of inertia: 'far enough away from another object'. And there is no reason to believe that the observed deviations from inertial motion are not due to previously unrecognized objects. Thus any such deviations are a powerful tool for detecting objects, previously unseen. More about this in Chapters 11ff.

D. Systematic Motions

We briefly summarize the several systematic features in stellar motions, which have played a major role in understanding the state of motion of the galactic system. In 1783 Friedrich Wilhelm Herschel (1738–1822) discovered *solar motion* i.e., the motion of our Sun, reflected as a systematic 'secular parallactic' effect in the proper motions of a small number of stars, thus supporting the analogous nature of stars and the Sun, 'our' star. More than a century later, from radial velocities the amount of the solar velocity with respect to our nearest stellar neighborhood, our travelling companions in the galactic system (Chapter 1), was found to be close to 20 km s^{-1}.

In 1904 Jacobus Cornelius Kapteyn (1851–1922) announced the discovery of two *star streams*, reworded and reinterpreted by Karl Schwarzschild (1873–1916) as *ellipsoidal distribution*. This phenomenon is frequently referred to as *preferential motion*, the greater mobility of stars along an axis roughly pointing toward the constellations Sagittarius and Auriga. A survey of Kapteyn's two streams and the various analytical interpretations are given by Arthur S. Eddington (1882–1944) in Chapters 6 and 7 of his excellent readable book *Stellar Movements and the Structure of the Universe* (1914, MacMillan and Co., Ltd., St. Martin's Street, London).

Gustav B. Strömberg (1882–1962) stressed the phenomenon of *asymmetry* from

Cygnus to Argo, the increase of *group motion* with increasing dispersion in preferential motion for different groups of objects. Group motion is measured in the direction of Argo, and is equal and opposite (reflexion of) to the galactic motion toward Cygnus.

Both preferential motion and asymmetry perpendicular thereto, were successfully explained by the theory of *galactic rotation* developed 1926–1927 by Bertil Lindblad (1895–1965). Analogy with asteroid motions observed from the Earth, i.e. a differential rotation around the Sun was published in 1871 by J. A. H. Gyldén (1841–1896) and later studied by others. In 1927 Jan Hendrik Oort (1901–   ) discovered and formulated the *differential galactic rotation*, and earlier (1922) had made a study of *high-velocity stars*. The latter do not exceed the 'velocity of escape' and thus clearly exhibit asymmetry through their avoidance of the general 'forward' galactic direction toward Cygnus.

For details see *Basic Astronomy*, Chapters 24, 27, 28, 29, Random House 1952.

CHAPTER 3

STELLAR DISTANCES

A. Heliocentric Parallax. Stellar Aberration

Although the heliocentric viewpoint is of such overwhelming simplicity and beauty that from a purely descriptive standpoint its acceptance could hardly be challenged; the question arose whether or not additional evidence exists. If it is not the Sun which revolves around the Earth, but rather the Earth which revolves around the Sun, then the effect of the Earth's orbital motion should be revealed in the stars – just as it is the case of the planets in our solar system – in the form of parallactic shift in direction in the course of a year. The distance to the Sun has been measured in a variety of ways and was found to be very close to 150 000 000 km, i.e. the orbit of the Earth around the Sun would have a diameter of 300 000 000 km. So huge a baseline would seem to hold some promise of revealing annual parallactic changes in the positions of the 'fixed' stars. These shifts were not observed for three centuries after Copernicus; and in the meantime the heliocentric viewpoint was not acceptable to many astronomers, Tucho Brahe (1546–1601), for instance.

We now know that the early failure on the part of astronomers to notice any heliocentric parallactic shifts was due to the tremendous distances of the stars. It was not until 1838 that reliable observations of such a shift were made for one of the nearer stars; and only comparatively recently, after 1900, has the technique been developed which is adequate for an all-round attack on this problem. We recall that the Copernican view (1543) of the solar system was proposed long before the invention of the telescope (1608). While the older astronomers were able to measure directions and angles with an accuracy of one minute of arc at best (the angle subtended by one meter at a distance of three and a half kilometers), the use of the telescope suggested a much higher accuracy well below the arcsecond.

Several early telescopic attempts were made to detect the heliocentric parallactic effect for fixed stars, but all without success. However, one of these historical investigations in 1727 proved to be of great interest. The astronomer James Bradley (1693–1762) had a suitable telescope at his disposal at the Greenwich Observatory, and made another attempt to look for the annual parallactic effect. To create the most favorable conditions he observed a star, Gamma Draconis, which was in a direction almost perpendicular to the orbit plane of the Earth. Assuming that the star has no motion of its own, and that the Earth describes an orbit around the Sun a telescope would in the course of the year have to be tilted ever so slightly 'inward', i.e. toward the direction of the star as seen from the Sun. To his surprise, Bradley did not find any such effect, but discovered another type of tilt instead. In the course of one year his telescope had to be tilted slightly *forward*, i.e., toward the direction of the Earth's path at any particular moment. Bradley correctly ascribed the effect, called *stellar aberration*, to the finite velocity of light, which had been discovered in 1675 by Olaus Römer (1644–1710).

The position of a star as observed with a telescope is defined by the axis of the cone of light which converges from the objective toward the focal image of the star; this position may be rigorously measured by pointing the telescope so that the focal image of the star is bisected by a set of crosswires in the focal plane. Since the light rays have a *finite* velocity, it takes time for the light to travel from the objective toward the focus. Imagine now the telescope will have to be tilted slightly forward toward the direction of motion in order that the star image may still be observed on the crosswires. Obviously the distance of the star does not enter these considereations; what counts is the direction of the star and the velocity of light and the velocity of the observer i.e. the Earth.

For analogy imagine a person walking in a downpour holding his umbrella in such a way as to provide maximum protection against the falling rain. The umbrella has to be tilted forward slightly in the direction of motion in order to anticipate the raindrops which are falling with a finite velocity. The amount of tilt varies directly with the speed of the observer's motion and inversely with the speed of the drops; the faster the drops come down, the smaller the tilt need be. While the tilt of the umbrella against the rain may have to be considerable, the analogous tilt observed for fast moving 'light drops', the so-called aberration of light, is quite small, $20''.5$ at most; however, such a quantity could be easily measured in Bradley's time. Although Bradley's discovery thus unexpectedly confirmed the heliocentric viewpoint, his observations failed to indicate any parallactic effect, which therefore was judged to be very small. The total amplitude of stellar aberration amounts to as much as $41''$. In 1728 he also discovered the *nutation*, a periodic precessional term with a total amplitude of $23''$ in a cycle of 18.6 years, primarily due to the Moon.

As we shall see shortly, the maximum total parallactic shift for the nearest known star (Alpha Centauri) is only $1''.5$. No wonder that such an effect, more than a hundred times as small as aberration or nutation, had to await the use of much more precise techniques.

B. Measurement of Stellar Parallax

The Earth's orbit furnishes the ideal practical baseline for obtaining precise stellar distances by purely geometric means. As a result of the revolution of the Earth around the Sun, a nearby star seen from the Earth, describes in the course of a year an apparent annual *parallactic* orbit on a virtually stationary background of distant stars, which always are abundantly available. The unforeshortened semi-axis major of this orbit, the heliocentric or *annual parallax*, represents the angular measure as seen at the location of the star, of *one astronomical unit* of length. An observer, located at the star, would see the unforeshortened radius of the Earth's orbit under the same angle p. The amount of this annual parallax p expressed in seconds of arc is a measure of the nearness of the star. The distance r is expressed in parsecs; one parsec is the distance at which one astronomical unit subtends an angle of one arc sec: $r = 1/p$. Hence one parsec is 206 265 AU or 3.258 lightyears. Thus a star with a parallax of $0''.20$ has a distance r of 5 parsecs or 16.3 lightyears.

This is the direct geometric method for measuring the distance of a star. There is no difficulty in distinguishing the star's parallactic orbit from its proper motion (Chapter 2), which moves the star uniformly along the sky. The nearer the star, the larger its parallax; beyond a distance of about 100 lightyears the parallax of a star is so small, that generally the attainable angular accuracy, rarely less than $0\overset{''}{.}005$, becomes inadequate. The value of a general study of the very *nearest* stars is due to the comparatively accurate distances obtained for these objects.

An accurate visual determination of annual stellar parallax for a nearby star (actually a double star), was made in 1838 by Friedrich Wilhelm Bessel (1784–1846). How do we know that a star is near to us? The best criterion for nearness proves to be the star's proper motion. Bessel had noted that a faint star in the constellation Cygnus, referred to as 61 Cygni, had an unusually large proper motion, amounting to as much as $5\overset{''}{.}2$ per year; at this rate the star traverses an angle represented by the Moon's diameter in about 360 years. Such a large motion should be a sign of comparative proximity in the same way was that the apparent rate of motion of a bird or an airplane across the sky provides a clue to its distance. True, the star or the bird may be moving close to the line-of-sight; it has been found however, that generallky the size of the apparent cross motion of a star is an effective criterion for its nearness. Bessel's studies were rewarded with success; a measurable parallactic displacement was found. Bessel used the *differential* method followed by later photographic observers, measuring the star on a background of reference stars with small angular separation from 61 Cygni. His instrument was a heliometer, a 'double image' micrometer made by dividing the telescope objective in two halves which can slide by each other. More accurate than the traditional micrometer, and virtually free from the source of errors (refraction, instrumental) in 'absolute' or fundamental positions, right ascension and declination (Chapter 1), he obtained a value of $0\overset{''}{.}33$ for the parallax of 61 Cygni, now superseded by the 'modern' photographic value $0\overset{''}{.}31$.

Observing parallactic shifts may be illustrated by the following analogy. Suppose a bird is seen projected against a distant background of trees or mountains. Let us first assume that the bird is 'standing still', as a hummingbird does. If the observer is located on a merry-go-round, his change of position at different times will result in his seeing the stationary bird projected on different parts of the stationary background. The observed back-and-forth motion of the bird is fully ascribed to the observer's motion, which thus is 'made visible'. The back-and-forth shift provides a measure for the distance of the bird, the angular amount of shift varying with the inverse of the distance; if one bird is twice as far away as another, the former will reveal only half the 'parallactic' shift of the latter. The problem is not changed essentially if the bird is not stationary but moves along with a uniform speed in a certain direction. In this case the back-and-forth motion is simply superimposed upon the forward shift of the bird and a simple arithmetical calculation permits a distinction between this continued forward shift and the back-and-forth parallactic shift.

For a moment it may seem that the now shattered illusion of 'fixed stars' has deprived us of any 'background' on which to observe these small parallactic shifts. The ancient astronomers found these 'fixed' stars an ideal background for studying the paths of the planets, and we still do. Relatively few stars show appreciable proper motions, and even fewer exhibit measurable parallactic shifts. There is an overwhelming majority of faint distant stars which are virtually 'fixed', and for all practical purposes these form an adequate background for studying the nearer stars; the effect of possible small shifts on the part of these background stars can be kept within bounds, and allowance made for such fluctuations. The proper motions and parallactic shifts of the nearer stars are thus revealed as small but measurable changes in direction on a readily available virtually fixed background.

C. Current Work

It was not until 1886 that Charles Pritchard (1808–1893) determined parallaxes by the photographic method. Followed by others, including such leading astronomers as Henry Norris Russell (1877–1957) and Jacobus Cornelius Kapteyn (1851–1922). The photographic method found its culmination in the classical astrometric investigations by Frank Schlesinger (1871–1943), the basis of all subsequent work in this field. In the beginning of this century Schlesinger determined photographic parallaxes with a heretofore unattained accuracy, using the long-focus visual refractor of the Yerkes Observatory, which has an aperture of 102 cm and a focal length of 1937 cm. This work was continued with the large refractors at several observatories: Allegheny, Cape of Good Hope, McCormick, Sproul, Van Vleck, the Yale station of Johannesburg and others. Except for Allegheny, Cape and Yale, all these are visual refractors, used in conjunction with a yellow ('minus blue') filter, thus providing the so-called *photovisual* technique.

The light-gathering and focusing qualities of a refractor are provided by the refraction of light through a lens or system of lenses. By using a combination of two lenses one may obtain partial correction for the difference in refraction of light of different colors. Such a visually corrected achromatic objective provides the best compromise method for the focusing of light in the yellow-green color range, to which the eye is most sensitive. Other refractors have been corrected for the conventional photographic color range, the blue. Special plates used in conjunction with the refractor, however, have proved particularly useful for positional work. Besides using such a telescope visually, one may place a photographic plate in the focus of the objective; and sharply defined photographs may be made, if a yellow or rather 'minus-blue' filter is used to eliminate the out-of-focus blue rays. This photovisual technique yields results which have certain advantages over conventional photographs taken in blue light. The reasons are partly atmospheric, blue light being more subject to scattering, and partly cosmic, because of the predominantly red color of our stellar neighborhood.

The photographic procedure creates an image of a limited portion of the sky on the photographic plate. The power of a large refractor lies, in the first place, in its great focal length, which portrays any configuration of stars on a large scale. For example, with the Sproul refractor of a focal length of 36 ft (1093 cm), the scale value in the focal plane amounts to $1'' = 0.053$ mm. Thus the image of the Moon in the focus is about 100 mm across and fills the greater part of the useful optical field; secondly, the generally large aperture permits the photographing of faint stars. Last, but not least, there is the remarkable precision of photographically recorded positions; small variations in the positions of star images on photographs taken at different times are readily measured with considerable accuracy.

The angular diameters of even the nearest and largest stars are so small (less than $0''.1$) that no photographed star image can reveal any information about the star's diameter. Any star seen through the telescope appears as a minute disk because of the very nature of light. The star image on a focal-plane photograph appears as a considerably larger disk due primarily to the photographic spreading effect of the emulsion. Both visual and photographic star images are, of course, affected by turbulence in the atmosphere. The sizes of photographic images are of no interest other than to provide a measure for the brightness of the star. The images are simply diluted representations of the locations of the pointlike stars; the centers of these images represent the positions of the stars.

The visual refractor continues to be well suited for precise astrometric measures. A current illustration is the Sproul refractor (aperture 61 cm, focal length 10.93 m, scale 1 mm = $18''.87$ or $1'' = 53$ microns in the focal plane), used with 13 × 18 cm Eastman Kodak 103aG plates and a minus blue (Schott OG–515) filter. A bandwidth of about 600 Å is obtained around the minimum focal length $\lambda 5607$. The effective wavelength ranges from about $\lambda 5480$ for a blue star (of spectral type A0) to $\lambda 5525$ for a red star (of spectral type M0). Most parallax work is still being done with refractors; reflecting telescopes used to be unstable because of their limited field and changes in the figure of the mirror due to temperature variations.

The situation has changed; witness the success of the powerful astrometric reflector (aperture 155 cm) a superior instrument in operation since 1964 at the Flagstaff station of the U.S. Naval Observatory. This reflector designed by K. Aa. Strand (1907–   ) with a fused silica primary mirror, has a scale of 1 mm = $13''.55$ corresponding to a focal length of 15.22 m (see *Stellar Paths*, 8–9, 62, 1981).

The diameter of a well-blackened star image ranges from about 1 to 2 sec of arc (50 to 100 μ). The relative position of two photographic star images on a plate may be obtained with an average error of one micron or about $\pm 0''.02$.

Parallax measurements are made on a background of three or more reference stars, of magnitude 10 or fainter; generally these stars are sufficiently distant to serve as a close approximation to a fixed background. A small correction ranging from about $0''.001$ to $0''.007$ and averaging about $+ 0''.003$, is required to reduce the measured 'relative' parallax to 'absolute'.

Although the diameters of star images range from about 0.05 mm up to 0.20 mm, experience has shown that the relative location of two stars can be found from the measured centers of their images with an average accuracy that is only a small fraction of the image size. From a plate taken with the Sproul telescope we may obtain the photographic location of a star on an appropriate background of several stars with an average uncertainty of only 0.001 mm (one micron), which is about 0.″02 in terms of direction. This is the angle subtended by one meter at a distance of ten thousand kilometers which gives us an idea of the power of the long-focus photographic method in measuring small shifts in direction. Accuracy may be increased by taking several exposures on several plates on anyone night, and uncertainty may thus be decreased to as little as 0.″01. This small amount represents the limit of attainable accuracy in photographic positions with present equipment and methods. The heliocentric parallax is slightly over 0.″03 at a distance of 100 lightyears, the approximate limit of distance penetration by the photographic method.

A conventional parallax determination based on some twenty or thirty plates, each with two or three exposures, extending over several years yields an average error about $\pm$ 0.″005 for the relative parallax. By increasing the observational material, and by multiple parallax determinations obtained at different observatories, higher accuracy may be reached. It has been customary for some time to extend series of plates over several decades for the purpose of studying orbital motion and mass ratio for double stars and for detecting and analyzing perturbations due to unseen companions. Improved parallax results have been the obvious byproduct; internal probable errors as low as $\pm$ 0.″001 have been obtained.

The high positional accuracy of long-focus photographic astrometry is the result of three contributing factors: long focal length with resulting large-scale portrayal in the focal plane; stability of the photographic emulsion on the plate, and precision measuring engine. The attainable accuracy is affected by atmospheric and optical effects, which are minimized by maintaining stability or the mounting of the optical parts, objective or mirror, by reducing the spectral bandwidth and by always observing each star in the same position of the telescope, preferably close to the meridian, and the differential nature of the techniques.

The positions of the parallax star and the reference stars are obtained using various types of measuring engines. Considerable gain in accuracy is reached through the new measuring machines such as the SAMM and SCAN machines at the U.S. Naval Observatory or the Grant two-screw machine now in operation at the Sproul Observatory, where the location of the star images is obtained by photoelectric scanning. The measurements of anyone parallax series generally are reduced by a linear transformation to a standard frame, an adopted background based on the reference stars. The units are metric (millimeters or microns).

D. Parallax Factors, Heliocentric and Barycentric

Dynamic acceleration effects of the galactic field on an isolated star are far too small to be observable at present; such a truly single star has a uniform rectilinear space motion. In practice, over a short time interval, this also may be said of the path of the component of a visual binary with a long period of the order of several decades or centuries: any nonlinear orbital effect over a few years would be negligible. However, a secular acceleration effect may have to be taken into account. For this and other details see *Stellar Paths*, Chapters 5 and 8.

The path of a single star is represented by the following equations of condition:

$$X = c_X + \mu_X t + q_X t^2 + \pi P_\alpha ,$$
$$Y = c_Y + \mu_Y t + q_Y t^2 + \pi P_\delta , \tag{3.1}$$

where X and Y represent the position of the central star in right ascension (reduced to a great circle) and declination, measured and reduced to the *standard frame* based on the reference stars. The standard frame holds for the equator of a convenient epoch, for example, the year 2000. The first three terms on the right-hand side of the equation represent the heliocentric path defined by the position c_X, c_Y at a zero epoch, the yearly proper motion μ_X, μ_Y and the quadratic time coefficient q_X, q_Y i.e. half the observed acceleration. The latter may be significant for the observations on nearby stars covering several decades. The time t is counted from a convenient epoch; the unit of time is the solar or Besselian year, which begins when the right ascension of the mean Sun is $280°$. The Besselian fraction of the year, τ, is listed for example, in the American Ephemeris and Nautical Almanac. The fourth term represents the parallactic displacement of the relative heliocentric or annual parallax π in the respective coordinates. The *heliocentric parallax factors* P_α and P_δ represent the projected fractional equatorial coefficients of the unforeshortened angular value of one astronomical unit at the location of the star.

The heliocentric parallax factors in right ascension (reduced to great circle measure) and declination, respectively are as follows:

$$P_\alpha = R \left(\cos \varepsilon \cos \alpha \sin \odot - \sin \alpha \cos \odot \right),$$
$$P_\delta = R \left[(\sin \varepsilon \cos \delta - \cos \varepsilon \sin \alpha \sin \delta) \sin \odot - \cos \alpha \sin \delta \cos \odot \right], \tag{3.2}$$

where R is the radius vector of the Earth's orbit expressed in astronomical units, $\varepsilon = 23°27'$ the obliquity of the ecliptic, and $\odot$ the Sun's true longitude; α and δ are the right ascension and declination of the star. The formulae for the parallax factors may be simplified by the following substitutions:

$$p = 0.9174 \cos \alpha, \qquad a = 0.3979 \cos \delta - 0.9174 \sin \alpha \sin \delta,$$
$$q = - \sin \alpha, \qquad b = - \cos \alpha \sin \delta \tag{3.3}$$

whence

$$P_\alpha = R(p \sin \odot + q \cos \odot),$$
$$P_\delta = R(a \sin \odot + b \cos \otimes). \tag{3.4}$$

The positions are oriented close to the equatorial coordinate system of, say, the year 2000. Hence the calculations of P_α, P_δ, the position α, δ is reduced to the equator and equinox of the year 2000; any appreciable effect of proper motion is applied up to the epoch of the observation. A precession correction of $-0\rlap{.}''838 \, (2000 - \text{epoch})$ is applied to the values of $\odot$, to refer them also to the equinox of the observations.

These heliocentric parallax factors refer to the parallactic displacements with respect to the Sun. To further reduce to the barycenter of the solar system we should consider the *perturbation* of the Sun caused by the planets. Thus the observed stellar path is freed from the reflex motion of the known perturbation. Expressed in astronomical units the amounts of these planetary perturbations are the product of the mean distance (semi-axis major) from the Sun times the mass of the planet relative to that of the Sun. The effects of the terrestrial planets and of Pluto are negligible. The principal contributor is Jupiter (50%), followed by Saturn, Neptune and Uranus in that order:

Planet	Mean distance from Sun in AU	Period in years	Inverse mass $\mathcal{M}_\odot^{-1}$	Semi-axis major of perturbation in AU
Jupiter	5.20	11.86	1047	0.0050
Saturn	9.54	29.46	3502	27
Uranus	19.19	84.01	22869	8
Neptune	30.07	164.78	19314	16

The barycentric parallax factors are discussed in *Stellar Paths*, Chapter 9.

The determination of 'heliocentric' parallax is hardly affected by the Sun's perturbation since the comparatively long periods of the small perturbation by the planets does not measurably interfere with the annual parallactic effect.

The values of μ_X, μ_Y, q_X, q_Y, and π, obtained from a general least squares solution, are *relative* to the dependence weighted mean background of reference stars. The parallax values π are reduced to *absolute* p by a small correction ranging from $0''.001$ to $0''.007$ and averaging about $0''.003$. The results are converted to arc sec.

CHAPTER 4

STELLAR PARAMETERS

A. Masses, Double Stars

Friedrich Wilhelm Herschel in 1802 noted the relative orbital motion for the two components of Castor and for other binary stars, the first revelation of Newton's law of gravitation outside our solar system. The components describe 'Keplerian' orbits around each other. If referred to a background of distant, virtually 'fixed' stars, we find that the two components describe Keplerian orbits around their center of mass the latter proves to have a uniform rectilinear motion, as single stars do. The orbit of the components around the center of mass differ 180° in phase; their scales are inversely proportional to the masses of the respective components. On a background of reference stars, the binary components move in non-rectilinear paths, which are the resultant of the uniform rectilinear motion of the center of mass and the orbital motion of the individual components.

Observations over a sufficient time interval of the motion of the two components relative to each other permit a calculation of the geometric characteristics of the relative orbit of the visual binary, angular size and shape, its orientation in space, as well as the period of orbital revolution. Knowledge of the distance to the binary system is required to convert the apparent angular dimensions into linear measure.

Multiple stars exist in which the space-time dimensions always have a hierarchic sequence, obviously cosmological demand required for long-range gravitational stability.

The great abundance of binary stars in our stellar neighborhood begs a cosmological explanation. Meanwhile these objects are obviously 'useful' in providing us with the virtually only means of determining stellar masses. As contrasted with an 'optical binary', an accidental chance angular proximity of two unrelated stars, which actually are at vastly different distances. A physical binary is a pair if stars sufficiently close together so that mutual gravitation maintain their relatinship, their belonging together, as time goes by; the components of the double star continuously 'fall around each other'.

In accordance with the law of gravitatin, the space and time dimensions of their combined mass are given by

$$\mathcal{M}_A + \mathcal{M}_B = \frac{4\pi^2}{G} \frac{a^3}{P^2}$$

(4.1)

in c.g.s. units, where the constant of gravitation is

$$G = 6.67 \times 10^{-8}.$$

Or, in practical units we find the so-called *harmonic-relation*

$$\mathcal{M}_A + \mathcal{M}_B = \frac{a^3}{P^2}.$$

(4.2)

Here $\mathcal{M}_A$ and $\mathcal{M}_B$ are the component masses in terms of the Sun's mass (2×10^{33} g), a is the mean distance between the components expressed in *astronomical units* (one astronomical unit = 149 600 000 km, which represents the mean Earth–Sun distance), P the *period* in years. *The linear value of a is found by dividing its observed angular value a'', by the parallax p''.* If more over, the center of gravity of the two objects is located by noting how the two stars fall around it, the individual masses may be calculated.

B. Luminosities

We are prepared now to distinguish between the apparent brightness of a star and its intrinsic brightness which may be calculated as soon as the distance of the star is known. A parallax determination provides the intrinsic brightness of a star, assuming the apparent brightness to be known.

The naked eye stars in order of decreasing apparent brightness are arranged according to *apparent magnitude*, ranging from one to six. According to the psychophysical law the *geometric* progression of *luminosity* corresponds to the *linear* progression of *magnitude*. This was more precisely put in 1856 by N. M. Pogson (1829–1891), the light received from a first magnitude star is 100 times that from a sixth magnitude; the *light-ratio* for a difference of one magnitude is $\sqrt[5]{100} = 2.512$ i.e. the logarithm of the light intensity varies with 0.4 times the magnitude.

The magnitude scale is extended for telescopic stars, the faintest photographed thus far are beyond apparent magnitude 25. In what follows magnitudes refer to the *visual* range of light to which the eye is sensitive; photovisual if measured photographically (Chapter 3), different results are obtained if other color ranges are considered. The apparent visual magnitude of the Sun is *minus* 26.72.

The apparent brightness of a star depends on two things: its intrinsic *lumiunosity* or 'candle' power and its distance from us. Two stars which appear equally bright to us may of course have quite different luminosities, the apparent equality being caused simply by difference in distance. A knowledge of the distances of these stars enables us to compare their intrinsic brightness by placing them, so to speak, at the same distance – the apparent brightness of a star can be translated into intrinsic luminosity whenever the star's distance is known. Our knowledge about stellar distances leaves no doubt that the tremendous discrepancy between the apparent luminosity of the Sun and that of the stars must be attributed to the great remoteness of the stars as compared with the relative nearness of the Sun.

Take the bright star Vega; the light that we receive from Vega is exceedingly weak compared with the Sun's illumination. It is possible to arrange photometric experiments which will give us the actual ratio of the illumination provided by the Sun and by Vega. It has thus been found that 5×10^{10} stars like Vega are needed to equal the apparent brightness of the Sun, while parallax measurements have revealed that the distance to Vega is $24\frac{1}{2}$ lightyears.

To illustrate a comparison of the intrinsic luminosity of Vega with that of the Sun, the distance to the Sun is 149 600 000 km; expressed in light time this distance is only

$8\frac{1}{3}$ lightminutes or $1/63\,300$ of a lightyear. Hence Vega is $24\frac{1}{2} \times 63\,300$ or about $1\,560\,000$ times as far away as the Sun. The dilution in the intensity of light from a source small in size compared with the distance found by the light rays is inversely proportional to the square of the distance. As the light 'front' reaches distances whose ratios are 1, 2, 3, etc. the light is spread, or diluted, over a surface whose ratios are 1, 4, 9, etc. If, therefore, Vega were placed at the distance of the Sun, its brilliance would increase $(1\,560\,000)^2$-fold, and it would be evident that Vega actually emits 2.4×10^{12} divided by 5×10^{10} or 48 times as much as the Sun. We say, therefore, that Vega has a luminosity of 48 (times that of the Sun).

Technically the intrinsic brightness or *absolute magnitude* is defined as the magnitude a star would appear to have if placed at a distance of 10 parsecs (parallax $0''.1$). The *absolute* magnitude M is obtained from the (absolute) parallex p and the *apparent* magnitude m through the relation

$$M = m + 5 + 5 \log p \tag{4.3}$$

or

$$M = m + 5 - 5 \log r \tag{4.4}$$

where r is the distance in parsecs (Chapter 3).

For the Sun $m = -26.72$, $r = 1/206\,265$ parsec. The absolute magnitude of the Sun is therefore

$$M = -26.72 + 5 + 26.57 = +4.85\,.$$

At a distance of 10 parsecs the Sun would appear a fifth magnitude star. The total light output of a star compared with the Sun, the so-called *luminosity L*, or 'sunpower' (rather than candle power!) is given by

$$\log L = 0.4\,(4.85 - M)\,. \tag{4.5}$$

To illustrate this, we compare the luminosities of three extreme stars:

	Vis. mag.	Parallax	Distance in lightyears	Abs. vis. mag.	Visual luminosity
Rigel	0.1	$0''.013$	250	-4.3	4600:
Procyon	0.4	0.292	11.2	2.7	7.2
Barnard's star	9.5	0.547	6.0	13.2	0.00046

Note the huge difference in visual luminosities of the two bright stars Rigel and Procyon at widely different distances. Compared with Procyon, Barnard's star, although twice as near to us, appears nine magnitudes fainter. Note the tremendous range in visual luminosities, 4600: for Rigel, 0.00046 for Barnard's star. In Chapter 6 we shall learn more about the range in luminosities, which for intrinsically faint stars is known to extend down to absolute magnitude 19, or a visual luminosity as low as 0.000002 times that of the Sun.

C. Spectra and Spectral Types

Stars differ in color, or what amounts to the same in surface temperature. The star's surfaces radiate in all colors, normally in color ranges or spectra which are continuous save for a delicate pattern of absorption lines caused by the star's atmosphere. The chemical composition of stellar atmosphere being essentially uniform, the pattern of absorption lines, as determined by their relative prominence, the so-called spectral type is primarily determined by the surface temperature.

The laws of spectral analysis have been applied with great success to Sun and stars. An analysis of the light from these objects reveals absorption spectra. According to Kirchhoff's law, the pattern of absorption lines indicates the presence of rarefied gaseous material between us and a source which yields the continuous background of colors in the spectrum. Like the Sun, any other star appears to have a tenuous gaseous atmosphere surrounding a somewhat hotter surface layer, the photosphere. We assume, therefore, that absorption lines are caused by atoms and molecules in a star's atmosphere which diminish the intensity of radiation in certain colors. A small fraction of the flow of energy is either absorbed or scattered by these atoms or molecules. The light from the objects has also passed through our terrestrial atmosphere, causing a number of *telluric* absorption lines due primarily to the oxygen and nitrogen in our atmosphere. Fortunately, these lines are not abundant enough to interfere seriously with the study of cosmic spectra. We can allow for their presence since it is known which lines are due to our own atmosphere.

Spectroscopic evidence emphasizes the all-important fact that the Sun and stars are very much alike. When studying stellar luminosities (Section B) and masses (Section A), they all led to the conclusion that, in a general way, the stars are suns and, conversely, that the Sun is a star. Both Sun and stars are self-luminous, consisting of comparatively dense incandescent material, surrounded by a cooler and rarefied atmosphere.

Several hundred thousand stellar spectra have been photographed; according to the pattern and intensity of the absorption lines they can be classified in a sequence. The most important is the Harvard classification developed by Annie Jump Cannon (1863–1941). With the aid of a prism placed in front of a photographic camera the images of all stars to about the ninth magnitude were drawn out into their respective spectral ranges, the appearance of which strongly suggested a classification. Miss Cannon recognized about a dozen principal spectral types which are now labeled by the letters O, B, A, F, G, K, M, R, N, S and a few others. Each spectral class has decimal subdivisions. Over 99% of all fell in the six groups B, A, F, G, K, M. All reveal absorption lines of hydrogen, which are strongest in type A. Only in types B and O do we observe faint lines of helium. As we proceed toward types F, G and beyond, we note a conspicuous general increase in the number and intensity of metallic lines. In types K and M absorption bands due to molecular compounds make their appearance.

The continuous background spectrum of the stellar surface corresponds closely to black-body radiation i.e. stars generally are perfect radiators.

The Sun's spectrum is of type G and is indistinguishable from the spectra of thousands of other G-type stars studied so far. The absorption of an A star like Sirius contains a series of conspicuous absorption lines attributed to the presence of hydrogen atoms in the star's atmosphere. The corresponding absorption lines of hydrogen in the solar spectrum are relatively weak. On the other hand, the Sun contains conspicuous absorption patterns due to all sorts of metals (obviously present in the solar atmosphere); in the spectrum of Sirius the corresponding patterns of metallic lines are extremely weak. Does the difference in spectrum imply that the atmosphere of Sirius has a different chemical composition from that of the Sun, in the sense that hydrogen is relatively more abundant for Sirius and the metals more abundant for the Sun? This is a possible, but not necessary complete explanation. It is not only a matter of what atoms are present in the atmosphere, but how many among those present participate in trapping certain kinds of light coming from the photosphere. Could it be that not all the atoms of any particular element are doing the same thing in different stellar atmospheres? Can it be possible that there are external conditions which determine the character and rate of the atom's activities in the respect?

D. Surface Temperatures and Colors

Surface conditions do differ from one star to another, the principal factor being temperature. Differences in surface temperature are indicated by differences in the distribution of the spectral intensities of different colors.

According to the laws of radiation the total amount of radiation over all colors increases with the temperature of the star, i.e. the higher the temperature of the star, the greater the intensity of total surface radiation.

For perfect radiators this total radiation per unit surface, the *surface brightness*, of a star is proportional to the fourth power of the absolute surface temperature. In formula:

$$E = 5.75 \times 10^{-5} T^4 \text{ ergs}^{-1} \text{ cm}^{-2} \text{ s}^{-1}. \tag{4.6}$$

The frequence of highest intensity being proportional to the absolute surface temperature. The spectral region of maximum intensity is shifted from the red to the violet as the surface temperature increases from lower to higher temperature

$$v_m = 1.04 \times 10'' T. \tag{4.7}$$

Since the *color* of a star, as judged by the eye, depends on the mixture of the different colors, the composite or *effective* color is found to shift from the red to the blue with increasing surface temperature. By studying either the amount of radiation in different colors, or the resulting effective color, the temperature of a star may be ascertained by means of the known laws of radiation. The frequency v_m of the color representing the highest energy is proportional to the surface temperature T: according to the formula by W. Wien (1864–1928):

$$T = 9.65 \times 10^{-12} v_m. \tag{4.7}$$

For example compare the white star Sirius (spectral type A) and the yellowish star, our Sun (spectral type G). The continuous background of the spectrum of Sirius is relatively more intense in the blue than is the Sun's. The maximum intensity of radiation in Sirius occurs in the ultraviolet color of frequency 1.10×10^{15}. Hence, the surface temperature of Sirius is 10 600 K, and that of the Sun (about) 5750 K.

The remarkable general result is the close relation between color or surface temperature and the pattern of the absorption lines of stars. The spectral sequence from O to S (Section C) is one of descending surface temperature. The Harvard sequence of spectral types proves to be a temperature sequence, the surface temperatures ranging from 25 000° for the bluish B stars to less than 3000° for red M stars. The corresponding atmospheric temperatures are only slightly lower, and play a major role in determining the absorbing activities of atoms and molecules in atmospheres which, with minor exceptions, have the same chemical composition for all stars.

The six principal spectral types are nicely represented by six of the brightest stars, all visible in the winter sky in the northern hemisphere, listed below:

Name	Vis. mag.	Spectrum	Color	Approximate surface temperature
Rigel	0.1	B8	bluish white	20 000 K
Sirius	− 1.5	A1	white	10 000
Procyon	0.4	F5	yellowish white	7 000
Capella	− 0.1	G	yellow	6 000
Pollux	1.1	K	orange	4 000
Betelgeuse	0.5	M1	red	3 000

A careful observer will recognize the color sequence which these six bright stars reveal to the eye, although the color-estimate may differ from one observation to another, depending on the color-sensitivity of the eye. As a matter of fact, this color reaction may differ for the two eyes of one and the same observer. The color in the table should therefore be considered a 'good try' and serve only as a guide. Interesting and exciting is it not, that a rough value of the surface temperature of a star can thus be obtained from a visual color estimate!

E. Diameters

A star's size or diameter may be derived by comparing its total intrinsic luminosity with its surface brightness, i.e., radiation per square centimeter. The intrinsic luminosity can be determined if both apparent brightness and distance are known (Section B). The surface brightness, including all frequencies, varies in proportion to the fourth power of the surface temperature; the latter may be determined from the observed distribution of radiant energy in the continuous background of the stellar spectra. In most astronomical studies the measured luminosities do not include the total radiation of the stars, since, like the eye, most instruments are set to receive a limited range of radiative frequencies. There is considerable practical difficulty in obtaining the total radiation comprising all frequency. Often it is advisable to limit the study to a certain range of radiation which can be conveniently observed with the available instrumental means.

Fortunately, the Earth's atmosphere is quite transparent for radiation in the visual range. Many astronomical studies are therefore based on the conventional radiation limited to the visual octave.

The *diameter* of a star may thus be found by comparing the star's luminosity with its surface brightness i.e. radiative temperature. The total radiation E of a star is proportional with the star's surface and to the fourth power of its temperature, i.e.

$$E \sim D^2 T^4 , \tag{4.9}$$

where D is the diameter of the star. Hence the star's diameter is proportional to

$$\frac{\sqrt{E}}{T^2} . \tag{4.10}$$

The size of a star is now derived as follows: We compare the star's surface brightness with that of the Sun, whose diameter (1 391 000 km) is well known from direct geometric measurements. Take for example, the bluish star Sirius; knowing its apparent brightness and distance (8.6 lightyears), we can deduce its visual luminosity, which is found to be 26 times that of the Sun in the visual octave of radiation. Its surface temperature is about 10 600 K compared with the surface temperature of the Sun, which is only 5750 K. The visual surface brightness of Sirius, i.e. its total visual radiation per square centimeter per second, is about eight times as great as that of the Sun. If Sirius had the same surface brightness as the Sun, it would therefore have 26 times the Sun's surface, or $\sqrt{26}$ times the Sun's diameter. Since, however, its visual surface brightness or radiative efficiency is eight times as great as that of the Sun, a surface of 26/8 or 3.25 that of the Sun will account for the luminosity of Sirius. Hence, we deduce that the diameter of Sirius is $\sqrt{3.25}$ or close to 1.8 times that of the Sun.

Take as another example a cool star such as the faint red star Krüger 60 B, which has a surface temperature of 3000° and a visual luminosity of 1/2250 that of the Sun. If the star had the same surface brightness as the Sun, its surface would be 1/2250 times that of the Sun. The surface brightness or efficiency of radiation in the visual octave at this temperature is, however, about 90 times as weak as the Sun's; hence, a surface of 90/2250 or 1/25 of the Sun's surface is required to account for the observed light and we conclude that the diameter of Krüger 60 B is $1/\sqrt{25}$, or one-fifth that of the Sun.

Other methods exist for the eclipsing binary character of certain stars.

CHAPTER 5

THE NEARBY STARS

A. The Brightest Stars

Before turning to the nearest stars, we shall first make another survey. From the observational point of view the very brightest stars offer obvious advantages. However, apparent brightness is not a good criteron of distance (Chapter 3). True, on the average the apparently brighter stars are nearer, but it is hopeless even to make a guess at the distance of individual stars on the basis of brightness alone. The reason is the tremendous range in stellar luminosities, illustrated by the limited sample of the brightest twenty-one stars, not including our Sun, arranged in order of their apparent visual brightness (Table 5.1). These stars are found in different parts of the sky. Several objects are double, multiple, faint companion stars have been ignored; Capella and Alpha Crucis are blended.

TABLE 5.1

The brightest twentyone stars

Name	m_v	Sp.	Parallax	Luminosity	
Sirius	− 1.46	A	0".378	26	*
Canopus	− 0.72	F	28	1 670 :	
Capella	− 0.08	G	80	150	*
Arcturus	− 0.04	K	97	100	
Alpha Centauri A	− 0.01	G	750	1.5	*
Vega	0.3	A	133	48	
Rigel	0.12	B	13	4 600 :	
Procyon	0.38	F	292	7.2	
Achernar	0.46	B	26	830 :	
Betelgeuse	0.50	M	5	22 000 ::	
Beta Centauri	0.61	B	9	6 300 ::	
Altair	0.77	A	202	11	
Aldebaran	0.85	K	54	140	
Antares	0.96	M	24	630 :	*
Spica	0.98	B	23	660 :	
Alpha Crucis	1.05	B	8	5 700 : :	*
Pollux	1.14	K	94	35	
Fomalhant	1.16	A	149	13	
Deneb	1.25	A	− 6	− : :	
Beta Crucis	1.25	B	2	− : :	
Regulus	1.35	B	0.045	130	*

Based on Hoffleit and Jaschek, Bright Star Catalogue, 1982.
* Double or multiple star.
: Uncertain.
:: Very uncertain.

The luminosity of each has been calculated from the apparent m_v and the trigonometric parallax p (Chapter 4). While these stars differ little in apparent brightness, they exhibit a great range in their distances and hence their luminosities, which are all intrinsically brighter than the Sun. The brightest seems to be Betelgeuse, one of the bright stars in the constellation Orion, which abounds in stars of high luminosity. The intrinsically faintest is the bright component A of the triple star Alpha Centauri, and is only 1.5 times as luminous as the Sun. It is the nearest of these apparently bright stars.

The bright stars yield data of considerable interest; they draw attention exclusively to the existence of intrinsically luminous stars but exclude the major portion of the stellar population (see C. de Jager, *The Brightest Stars*, D. Reidel Publ. Co.).

B. The Nearest Stars. Spacing of Stars

Now we shall limit ourselves to the stars which we know best, namely those in our immediate cosmic neighborhood, a minute fraction of a spiral arm of our Galaxy. By considerable effort reliable parallaxes have been determined up to distances of about 50 parsecs. Generally a limit of 20 parsecs appears indicated for trustworthy results. The nearest stars, within 22 parsecs, have been the subject of continuing study by Wilhelm Gliese.

Five parsecs often have been a conventional limit for our 'immediate' stellar neighborhood. We shall derive general, and specific, properties of the stellar population, by studying this most precise sample, namely the more than sixty stars known to be nearer than 17 lightyears or 5.2 parsecs. Studies extending beyond this arbitrary limit have essentially confirmed the properties derived from the more limited study; information becomes less precise with increasing distance. A study like the present contributes little to knowledge about the grand structure of the Milky Way System. It does, however, give rather precise information about individual properties of stars and on the interrelation between these properties, which may be expected to prevail along the spiral arms in the disk of our Galaxy.

The distance limit of 17 lightyears is arbitrary, but is enough to include a sufficient number of stars thus providing a workable sample of adequate scope. How have we found the stars within this particular limit of distance? We recall (Chapter 2) that a star's proper motion is by far the best criterion for its distance. For several decades astronomers have determined the distances of stars of large proper motion; and, with hardly any expectations, such stars have proven to be relatively near.

As of March 1985 the sample of stars within 5.2 parsecs contains 62 separate visible stars, including the Sun. Of this sample fewer than a dozen stars are visible to the unaided eye. The majority are faint telescopic objects which were first recognized as being close to us because of their large proper motion. Not all 62 stars are isolated like the Sun. Several prove to form binary or even triple systems whose components are relatively close together, compared with the general spacing between stars. As time goes by such stellar systems prevail, due to the strong gravitational bond between the components.

In this nearby sample, twenty-nine stars, including the Sun, appear single. We note that the Sun is generously endowed with planetary and other dependents.

Twenty-eight stars are grouped two by two in fourteen binary systems. Six stars are grouped three by three in two triple systems, which may be considered to consist of an 'inner' binary system and a distant companion. Within the admitted incompleteness of the material, more than one-half of all stars are components of double and triple systems. In addition, irregularities in the proper motions of several of the stars indicate the existence of unseen companions.

These nearby stars fill a volume of 589 cubic parsecs, corresponding to a density of 0.093 stars per pc^3, or 1 star per 330 cubic lightyear. If we consider 47 systems rather than separate stars we find a spatial density of 0.080 systems per pc^3, or about 1 system per 440 cubic lightyear. This corresponds to an average separation of somewhat over 2 pc, or about 8 lightyears. Compare these with stellar diameters, which generally are well below one lightminute; the diameter of the Sun is less than 5 lightseconds. The additional invisible stellar (and other) companions add to the space-density of the stellar population.

The spacing within the double and triple systems is extremely small compared with the wide spacing between the different single, double, and triple star systems. The separations between the components within binary and multiple systems are generally well below one light day but range from about one-fourth light day to one-fifth lightyear. Or in terms of the mean Earth–Sun distance, the astronomical unit, the spacings of stars within these systems range from several AU for the separation of the triple system of Alpha Centauri for the component C from the 'inner' double star AB.

The tremendous space between stars, the emptiness of our cosmic neighborhood, may be illustrated by a scale analogy of over some sixty small spheres – tennis balls, golf balls, marbles, and a large proportion of smaller objects spread at random through a spherical volume the size of the Earth, that is 12 740 km in diameter. Their motions on the average would be something like ten meters per year. In this analogy, while an extreme separation of about 75 km is reached within the Alpha Centauri system, the internal separation of the component stars in binary and triple systems is generally less than 1 km, a very small fraction of the average spacing of several thousand kilometers between the different objects.

These numerical conclusions are, of course, affected by incompleteness of the sample which is indicated through the observed decreasing value of the population density with increasing distance. The incompleteness of our knowledge about intrinsically faint stars, even those in our immediate neighborhood, is illustrated as follows: Suppose we count the stars up to a distance of 13.5 lightyears. The amount of space thus covered is one-half the larger sphere with the radius of 17 lightyears. The smaller sphere contains 25 objects, which is well over one-half the number in the larger sphere. The incompleteness is also revealed in the observed abundance of stellar components, even if we exclude the planetary companions of the Sun and the unseen companions of other stars. The twenty-five stars nearer than 13.5 lightyears represent thirty-six separate components; while the stars whose distances are between thirteen and a half and

17 lightyears include only 26 observed components. The observed state of duplicity or multiplicity decreases with the distance; the same is true for the number of unseen components. This decrease must be due to increased difficulty in observing the existing components.

The decrease in star density becomes still more striking if we extend the distance limits. It also appears to be essentially due to observational incompleteness of the overwhelming abundance of intrinsically very faint stars. Only four stars nearer than 17 lightyears are found to be more luminous than the Sun. At the top, at 8.6 lightyears, is the bright component of the binary star Sirius, the brightest star in the sky with a luminosity of 26 times the Sun. Next comes Altair at 16.5 lightyears, with a luminosity 11 times that of the Sun. The third brightest is Procyon A, the brightest component of the double star, at 11.4 lightyears, with a luminosity of 7 times the Sun. Alpha Centauri A at 1.4 lightyears, the brightest component of the triple system, appears to be slightly more luminous (1.5) than the Sun. The intrinsically faintest star in the sample is Wolf 359; this star is only 1/50 000 as luminous as the Sun. Although it is the third nearest star (excluding the Sun), it is fainter than the thirteenth apparent visual magnitude, and difficult to observe, both visually and photographically. Outside our sample, at a distance of 19 lightyears, is Van Biesbroeck's star – the star of lowest known luminosity, 10 times as faint as Wolf 359. We are prepared for the discovery of still fainter stars within the limits of the present sample.

The discovery and subsequent study of these low-luminosity objects, have been accelerated through the photographic propermotion surveys of apparently very faint stars by Willem J. Luyten (1899–   ) and others. As mentioned earlier, apparently faint stars of appreciable cross motion on the sky have almost always proven to be comparatively nearby and of low intrinsic luminosity. For every star intrinsically brighter than the Sun there appear to be over a dozen which are less luminous than the Sun. The overwhelming majority of the population of the Universe, as revealed in our immediate cosmic neighborhood, consists of stars of very low luminosity. This is illustrated in Table 5.2, which gives the visual luminosity function i.e. frequency distribution of luminosities. The sharp drop below 0.0001 times the Sun's luminosity must be partly due to incompleteness of the data; it should gradually change through future parallax measures of numerous faint proper motion stars waiting observation.

TABLE 5.2

Visual luminosity in terms of the Sun	Number of stars
1 – 23	4
0.11 –1.0	7
0.011 –0.10	6
0.0011 –0.010	7
0.000 11–0.001 0	20
0.000 02–0.000 10	6

In contrast with the five stars intrinsically as bright or brighter than the Sun, there are fifty-eight stars intrinsically fainter than the Sun; the combined visual radiation from these stars reaches just about twice the Sun's luminosity.

C. Main Sequence; Red and White Dwarfs

We are in a position to make further studies of the nearby stars in so far as they refer to luminosities, colors (spectra) and sizes. With a few exceptions, there proves to be a remarkable relation between luminosities and spectra (colors). The fainter a star, the redder its color, so that an arrangement of the nearer stars according to decreasing luminosity also proves to be a sequence of an increasing redness, and of decreasing diameter. The astronomer refers to his color-luminosity relation as the *Main Sequence*. The stars on the Main Sequence are referred to as *dwarf* stars. The designation giant stars is used for intrinsically very bright stars, both red and blue, which are scarce and none of which appear in our immediate neighborhood. A striking result is the overwhelming majority of red dwarf stars, much smaller and fainter than the Sun.

Exceptions to the Main Sequence are certain stars that are small and of low luminosity, but not red. There are a few such stars in our sample. They are named *white dwarfs* or *degenerate stars*. The best known of these is Sirius B, whose luminosity is 0.028 times that of the Sun; its mass is very close to that of the Sun, however. The white color of Sirius B implies a surface temperature of about 10 000 K. That its total luminosity is so small implies that it is also small in size, less than 1/100 of the Sun's diameter. The density of Sirius B appears to be over 2 million times the density of the Sun, or something like 3 million times that of water.

Studies beyond the five-parsec limit have confirmed the spectrum (color) – luminosity relation, and confirm the existence of the Main Sequence and of the degenerate branch in our wider stellar neighborhood. We may state that in a wide cosmic neighborhood about 90% of all stars are 'regular' (Main-Sequence) dwarf stars and about 10% are

TABLE 5.3

Number of stars of different colors (spectra) nearer than 17 lightyears

Color	Spectral type	Number of stars
White	A0–F5	3
Yellow	G0–G8	3
Orange	K0–K7	9
Red	M0–M8	42
Total Main Sequence		57
White dwarfs		5
Total		62

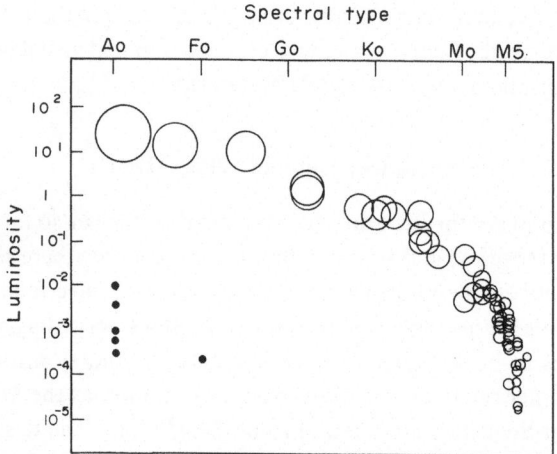

Spectrum–luminosity relation for stars within 17 lightyears. Spectra and visual luminosities for stars nearer than 17 lightyears. Approximate relative diameters are indicated by open circles except of the white dwarfs (dots).

white dwarfs. Less than 1% are the intrinsically very bright giant and supergiant stars. As to objects fainter than the extremely faint red dwarfs, they will be considered in Chapter 6.

Stars are spheres of glowing gases, primarily hydrogen, which provide the star's radiant energy by nuclear reactions in the interior, primarily and initially the conversion of hydrogen into helium. The Main Sequence is explained through the formation of stars from the contraction of spheres of interstellar material. The place of a star on the Main Sequence is essentially determined by the amount of matter in the original spheres; large masses lead to highly luminous stars, the obviusly very abundant stellar spheres lead to the large number of stars of low luminosity. Theory suggests that objects with a mass less than 6% of the Sun no longer derive their radiant energy from nuclear reactions.

The very fact that the majority of stars in our immediate neighborhood follow this narrow Main-Sequence pattern, is an indication of the 'comparative youth' of these stars. The exceptions are the few white dwarfs, which have moved off the Main Sequence. The white dwarfs are explained as former Main-Sequence stars, which exhausted energy production by conventional nuclear reactions. These stars are collapsed and still contracting; their atoms are crushed, and referred to as degenerate matter. These high-density objects are stars that 'have seen better days' and are now in the last phase of their lifetimes.

D. Mass-Luminosity Relation. Mass Density

In our sample there are a dozen binaries for which good orbital data exist; the nearby stars furnish some of the best determined stellar masses that we know. Sirius A is found to have slightly over twice the Sun's mass; the smallest well determined mass – 0.06

that of the Sun – belongs to the faint red dwarf Ross 614B. Beyond the limits of the present sample, some stars with masses up to 50 times that of the Sun or more have been found. No objects with mass below 0.06 that of the Sun have been seen, although their existence is indicated by their gravitational effect on 'parent' stars. There appears to be a well-established general relation between mass and luminosity for the Main-Sequence stars. The higher the mass, the higher the luminosity; however, the range in mass is small compared with that in luminosity, the luminosities vary approximately with the third power of the mass. Exceptions to the Main Sequence and the general luminosity relation are the white dwarfs, which are thought to be former Main-Sequence stars, having substantial masses. Masses for three white dwarfs are known: Sirius B $(0.94\ M_\odot)$, Procyon B $(0.65\ M_\odot)$, and 40 Eridani B $(0.43\ M_\odot)$. Theoretical considerations explain both the mass-luminosity relation and the exceptions to it.

While a decade and a half ago the lower end of the mass-luminosity relation was not known beyond Krüger 60B, this relation has now been extended through the data for Ross 614. Whether masses lower than $0.06\ M_\odot$ exist for true stars is a challenge to theoreticians and, we trust, will be settled by observers.

The mass-luminosity relation permits us to estimate the masses of the other single Main-Sequence stars; the masses of the three single white dwarfs in this sample are estimated at $0.6\ M_\odot$ each. In this way a total mass of twenty-five solar masses is found for the (visible) stars in the present sample. This leads to an average density of 0.042 per pc^3, or an average mass-density of nearly $3 \times 10^{-24}\ g\ cm^{-3}$. This value is close to the estimated density $3 \times 10^{-24}\ g\ cm^{-3}$ of interstellar material in our neighborhood. (Compare this virtual vacuum with the density $10^{-3}\ g\ cm^{-3}$ for air and $10^{-14}\ g\ cm^{-3}$ for the best vacuum obtainable in the laboratory.) Relative to the interstellar material, stars obviously are condensations of a comparatively much higher average density; be it only $10^{-8}\ g\ cm^{-3}$ for red super-giants such as Betelgeuse, $10\ g\ cm^{-3}$ or more for red dwarfs, or $10^5\ g\ cm^{-3}$ or more for the white dwarfs.

CHAPTER 6

THE LOWER END OF THE MAIN SEQUENCE

A. The Intrinsically Faintest Known Visible Stars

How far down does the Main Sequence go? Is there a lower limit? Is there a gradual or perhaps a sudden change from stars to planets? We shall consider these questions both from the observational and theoretical side.

The next nearest star, distance 1.83 parsec, Barnard's star, an M5 dwarf of apparent visual magnitude 9.54, has an absolute visual magnitude: + 13.2. The faintest component C (Proxima) of the nearest stellar system Alpha Centauri, at a distance of 1.33 parsecs, apparent magnitude 11.0, spectrum M5e, has an even fainter absolute visual magnitude: $M_v = + 15.4$. Within 5.3 parsecs several more stars are found, all fainter than $M_v = + 15$. All these are listed in Table 6.1.

TABLE 6.1

Stars nearer than 5.3 parsecs with M_v fainter than + 15

Name	R.A.	Decl.	p	m_v	Sp.	M_v
Proxima	14^h29^m6	− 62°40′	0″.762	11.0	M5e	+ 15.4
Wolf 359	10^h55^m5	+ 7° 0′.9	0″.419	13.5	dM8e	+ 16.6
L726−8 A	1 39.0	− 17 57.0	0.382	12.5	M6Ve	15.4
B				13.0	M6Ve	15.9
G51−15	8 29.8	+ 26 46.6	0.276	14.8	M6.5Ve	17.0
Ross 614 B	6 29.4	− 2 48.8	0.251	14.8		16.8
Wolf 424 A	12 33.3	+ 9 1.3	0.228	13.1	M5Ve	14.9
B				13.4		15.2
G158 27	0 6.7	− 7 32.4	0.214	13.7	M5–5.5V	15.4
G208−44	19 53.9	+ 44 25.4	0.212	13.4	M5.5Ve	15.0
45				14.0		15.6
G9−38 A	8 58.2	+ 19 45.7	0.192	14.1	M8Ve	15.5
B				14.9		16.3

An obvius and effective search for intrinsically faint stars in our neighborhood was made by George Van Biesbroeck (1880–1974). From ten-minute exposure plates taken with the 208 cm reflector at the MacDonald observatory he found numerous faint distant 'proper motion companions' of known nearby stars. A time interval of only a few years was needed to reveal such objects of at least the eighteenth magnitude through the proper motion they shared with the central star of known parallax. In this simple manner Van Biesbroeck found a dozen stars of low luminosity, labelled VB1 to VB12 (*Astron. J.* **66**, 528–530, 1961). In addition he found 17 stars of appreciable proper motion, labelled VB13–VB29, and of interest for parallax determinations.

Of these the intrinsically faintest object is VB10, also known as 'Van Biesbroeck's star', the distant companion of BD + 4°4048, apparent visual magnitude 17.4, at a distance of 5.8 parsecs, with an absolute magnitude $M_v = + 18.6$. We shall have more to say about VB10, also VB8, in Chapter 13. Both stars are of particular astrometric interest, since they appear to have perturbations.

A still fainter object has been found by I. Neill Reid and Gerald Gilmore (*Monthly Notices*, July 1981; report in *Nature*, Sep. 3, 1981 by David W. Hughes). This star is RGO 050–2722, at a distance of about 25 parsecs and apparent visual magnitude about 20. The distance was determined photometrically; the color is consistent with a surface temperature of 2625 K. Its (estimated) absolute magnitude is $M_v = + 19$ or fainter. In Table 6.2 are listed the 4 stars with lowest known absolute visual magnitude fainter than 17.0.

TABLE 6.2

Stars with M_v fainter than $+ 15$

Name	R.A.	Decl.	p	m_v	Sp.	M_v
VB8	16^h55^m6	$- 8°23'.6$	$0''.160$	16.8	M7V	17.8
VB10	19 17.0	$+5$ 8.8	0.178	17.4	M8Ve	18.6
RGO 050–2722	0 52.9	$- 27$ 6.0	0.04 :	20 :		19 :
L 271–25	14 28.7	$+ 33$ 10.6	0.115	19.7	dM9	20.0

It is very likely that many more intrinsically faint stars remain to be discovered in the immediate stellar neighborhood. Statistical studies of the frequency of intrinsically faint stars by Luyten and others indicate a maximum of about $M_v = + 14$ followed by a relatively steep decline. Any definite limit is not yet established; there is no reason to believe that stars with absolute visual magnitude $M_v = + 20$ or even fainter may not exist. In any case theory of stellar structure and evolution permits relatively precise assertions.

Minimum star size is governed by the size of the smallest protostar cloud fragments which are stable against disruptive forces. This probably depends on the rate of loss of angular momentum, the opacity of the fragments and the physical properties of the original interstellar cloud. According to theoretical investigations a *lower mass limit* for the formation of single star-like objects through fragmentation of gravitational unstable interstellar clouds is to be expected between $0.01 \, \mathcal{M}_\odot$ and $0.005 \, \mathcal{M}_\odot$.

Another question is, what is the *minimum mass* required for the formation of a 'normal' Main-Sequence star? Stars result from the contraction of large cold spheres of gas (predominantly hydrogen). Contraction leads to increased temperature, until a stable star results, namely a gaseous sphere, in which the equilibrium between gravitational pressure and outward gas and radiation pressure determines the central temperature. The smaller the mass of the star, the lower the central temperature and the lower the luminosity of the star. The zero-age Main Sequence is thus explained down to its lower section: the red dwarfs. There is however a critical mass value below which the central temperature is too small to permit the conventional nuclear energy production by the conversion from hydrogen into helium, and no more Main-Sequence stars occur. The resulting objects are no longer red dwarfs stars and are called *substellar*, *black*, *brown* or *even dark red stars*. I discussed this matter of nomenclature in June 1982 with Martin Schwarzschild. He indicated that 'substellar' would seem to imply a distinction from real stars. This Schwarzschild would agree to regarding their energy source (but are white dwarfs then also substellar?), but not regarding their mass distribution which presumably represents just the tail end of the mass distribution of ordinary stars. Schwarzschild proposed the name *dark dwarfs*, which I shall use from here on.

B. Red and Dark Dwarfs

We shall briefly anticipate and utilize some of the results presented in Chapter 8 on the problem of origins.

Dark dwarfs therefore start their lives just like red dwarfs, on a relatively fast Hayashi track contraction toward an extended region beyond the end of the red dwarf portion of the Main Sequence. Because of their low masses they are unable to build up central temperature high enough to ignite and lead to stable hydrogen burning. After a while, in the interior of the protostar, deviations from the ideal gas law occur. The gas becomes degenerate and the now valid equation of state results in a sphere for which the central temperature and also pressure are not sufficient to yield thermonuclear energy; from there on thermal energy is radiated. The radii *remain essentially constant* while the stars gradually grow *cooler* and fainter, much like a hydrogen-rich white dwarf. Eventually they become so small that they can be detected only in the infrared. While the Hayashi contraction phase is relatively short, evolution in the degenerate phase is slow. The relative number of those bjects is greatest at long wavelengths.

Calculations about the distinction between red and dark dwarfs have been frequently made. According to S. Kumar (*Astrophys. Space Sci.* **17**, 219, 1972) dark dwarfs may be defined as objects with very low masses ranging from 0.01 to 0.10 $\mathcal{M}_{\odot}$. Low and Lynden-Bell (1970), Scalo (1978) calculate a limiting mass range of 0.007 to 0.085 $M_{\odot}$ for dark dwarfs; red dwarfs are taken to have a range of 0.085 to 0.5 $\mathcal{M}_{\odot}$.

A. S. Graboske and H. C. Grossman, Jr. (1971, *Astrophys. J.* **170**, 163, 1974, *Astron. Astrophys.* **30**, 195) have calculated the evolution of protostellar gaseous spheres with masses 0.2 $\mathcal{M}_{\odot}$ and smaller (about $M_v > 13$ and spectral types M5V and beyond). They assumed the chemical composition of the interstellar medium and of the atmospheres of young stars: $Y = 0.29$, $Z = 0.03$. For a gaseous sphere of 0.08 $\mathcal{M}_{\odot}$ the ideal law does not permit sufficient energy production to counteract gravitational contraction. The lower limit for the mass of a stable Main-Sequence star on the edge of degeneracy appears to lie near 0.085 $\mathcal{M}_{\odot}$. For this value the calculations still manage to lead to a stable star. At this critical mass value the approach from above toward the Main Sequence (Hyashi track), lasts at most about 10^9 yr, and the resulting star has a luminosity somewhat below $1/1000\ L_{\odot}$ (bolometric). The corresponding visual absolute magnitude corrected for adopted color or temperature, is between 17 and 18 (if Van

Biesbroeck's star is a Main-Sequence star, its mass should be around $0.085 \, \mathcal{M}_\odot$ (edge of degeneracy).

The transition from the Hayashi to the cooling track is a most interesting part of the evolving track for dark dwarfs. On this part a substantial number of dark dwarfs should be visually detectable with about $M_v = +18$ or fainter, due to a combination of relatively high luminosity and long lifetime; traces of deuterium in the original cosmic material create some energy from the ${}^2\mathrm{H}(p\gamma){}^3\mathrm{He}$ reaction. At the top of the Hayashi track the star evolves too rapidly to be seen, at the lower end of the cooling track in the course of several 10^9 yr their temperatures and luminosities decrease so much that they will be too faint to be seen. They can only be detected at infrared wavelengths: they truly become 'invisible' dark dwarfs.

In order to have one and the same name for the same object in different stages of development, the name dark dwarfs is extended to their early visible phase. Briefly therefore we speak of red and dark dwarfs, the distinction being whether their mass is either more, or less than $0.085 \, \mathcal{M}_\odot$. At the lower end of the Main Sequence we recognize therefore

> hydrogen burning dwarfs: *red dwarfs*
> deuterium burning dwarfs: *visible dark dwarfs*
> followed by
> cooling track: *invisible dark dwarfs*

C. Statistical Studies of Dark Dwarfs

R. F. Staller and T. de Jong, under certain assumptions of stellar evolution and the mass spectrum of the created stars have calculated the luminosity function of red and dark dwarfs. The *mass spectrum* i.e. histogram of stellar masses

$$n(M) \, \mathrm{d}M^{-\alpha}$$

where α may be taken as -2.5 is based on
  (1) open clusters down to $>1 \, \mathcal{M}_\odot$ (NGC 2264)
  (2) stellar neighborhood: $0.1-1.0 \, \mathcal{M}_\odot$.
Staller and de Jong assume a stellar birth rate down to the limit $0.007 \, \mathcal{M}_\odot$ which remains constant throughout the life of the Galaxy. The red dwarfs will exist essentially unchanged for a time *longer* than the current life time of a galaxy.

For objects below the critical limit of $0.085 \, \mathcal{M}_\odot$ i.e. the dark dwarfs, at any one time there should be a substantial number visible with about $M_v = +18$ and fainter. In the course of several 10^9 yr their luminosities and temperatures however decrease so much that they can only be detected at infrared wavelengths, and they truly become 'invisible' dark dwarfs.

The following general conclusions may be drawn. The age of our Galaxy is some ten times larger than the time which a (future) dark dwarf spends in the luminosity range of the faintest red dwarfs. Assuming constant star creation in the past there should therefore be about 10 times as many invisible dark dwarfs as still luminous with masses below $0.085\,\mathscr{M}_\odot$. Unfortunately the true frequency of still luminous objects with absolute magnitudes 16 and weaker is poorly known. Staller and de Jong conclude that beyond the maximum of the general luminosity function, beyond about $M_v = 16$ or 17, essentially only future dark dwarfs appear. Hence the lower limit of the visual Main Sequence could lie between $M_v = 16$ or 17. They expect about 10 dark dwarfs scattered among one million observable infrared sources. The Infrared Astronomical Satellite (IRAS) is not considered promising for this problem; the situation is better with the Space Telescope. Assuming a limiting visual magnitude of $+ 26$, about 2000 dwarfs per square degree are expected, about 90 of them being dark. These estimates are quite uncertain.

D. Visible and Invisible Dark Dwarfs

T. van der Linden (1982) proposes five possible visible candidates for *contracting* dark dwarfs: the afore mentioned Wolf 359, VB 3, VB 8, VB 10, and RGO 050–2722.

VB 10 (van Biesbroeck's star) and RGO 050–2722 especially are strong candidates because they are 2 magnitudes fainter than the limit for stable hydrogen burning red dwarf stars ($M_v = + 17$).

From the relative number of visible dark dwarfs as a wide companion to a Main-Sequence star, Van der Linden estimates the lifetimes of these objects. If we assume 10^{10} *yr as the average age for the primaries*, that stars are continuously formed and *the dark dwarfs simultaneously as a companion*, then the average age of a black dwarf in terms of the average age of the primary is given by the number of observed dark dwarfs in terms to the total number of primaries.

Within 22 parsecs 4 dark dwarfs are found, and systems and single stars are observed, i.e. a ratio of 1/400. An average age of about 25×10^6 yr and 'mass of $0.02\,\mathscr{M}_\odot$' for dark dwarfs are thus found. The maximum lifetime for a $0.08\,\mathscr{M}_\odot$ object would be 10^8 yr. Conclusion: At least 1 out of 4 stars is accompanied by a dark dwarf.

E. General Remarks

(1) Unseen stars including dark dwarfs were discovered from *perturbations*, thus far with periods well below 100 yr.

(2) Very faint objects, such as the distant van Biesbroeck companions of nearby stars, by theory may be explained as dark dwarfs.

(3) Note that the extreme discovery methods (1) and (2) both yield dark dwarfs as *companions* of visible stars, the latter mostly being red dwarfs.

(4) Speckle interferometry in infrared is an effective method for finding 'bright' dark dwarfs, and perhaps not so bright ones.

The evolution of Low Mass stars through mass loss by transition from the Main Sequence to degenerate phase has been studied by F. D. Antona and I. Mazzitelli (*Astron. Astrophys.* **113**, 303, 1982). They suggest that some stars close to the Main Sequence lower mass limit may lose a significant fraction of their mass during the Galaxy lifetime and die as dark dwarfs.

CHAPTER 7

OUR SOLAR SYSTEM

A. Introduction

The principal intent of this treatise is to review our current state of knowledge of 'planetary and dwarf companions of stars'. It is reasonable to examine the companions of our own star, the Sun, the nearest planetary companions, which over the centuries have provided us with much information, the earliest and not the least of which, is the law of universal gravitation. Over the past decades we have been able to supplement our passive receiving of cosmic messages on the Earth's surface by the more aggressive approach of communicating with various planets in our solar system. This 'space' research, provincial though it may seem on a cosmic scale, has given us important and novel information about the physical and chemical composition of the planets and has contributed to better understanding of their origin (Chapter 8).

We take the planets in our solar system so for granted, that we might consider them as a unique phenomen, they could be – in terms of knowledge – they are. But we must remember that 'our' planets are dependents of 'our' Sun, and that there are countless other stars, many identical in structure and composition to our Sun, others either fainter or brighter (Chapter 5). Even for the nearest star it remains a difficult problem to detect their planets, if they have any. But the selection effect imposed on us by our location, and limited possibilities, should be no reason not to recken with the possibility that stars, other than our 'own' star, the Sun, may also be endowed with planetary companions.

That most stars are not alone is revealed by the observed fact that the majority of stars are not single, but exist as components in double and multiple stars. And gradually more and more invisible unseen, dark, companions are found for stars, previously considered single and we shall return to these problems in Chapters 11 ff. First let us survey the planets of our own Sun, i.e. our own planetary system or, if we wish, our own solar system, if we include other dependents as well, such as comets and meteors. Then in Chapter 8, we shall consider the possible origin and evolution of our solar system. And we should of course realize that any theory or hypothesis of the origin of our solar system, the Sun and its dependents, may be equally applicable to any other star.

B. Geometric and Kinematic Properties of Planetary Motions

Table 7.1 gives relevant information of some geometric and dynamical properties of the planets.

We distinguish at first between the Earth-like (terrestrial) planets, or *Earth-planets*: Mercury, Venus, Earth, Mars, and the Jupiter-like, (Jovian or Giant) or *Jupiter-planets*: Jupiter, Saturn, Uranus, Neptune. We have also listed the 'average' asteroids, and the 'abnormal' distant planet Pluto, which may well be a former satellite of Neptune,

TABLE 7.1

Name	Semi-axis major in AU	Eccentricity	Inclination to ecliptic	Period in years	Mass relative to Earth
Mercury	0.39	0.206	7°0	0.24	0.055
Venus	0.72	0.007	3.4	0.62	0.815
Earth	1.00	0.017	0.0	1.00	1
Mars	1.52	0.093	1.8	1.88	0.107
Asteroids	2.8 :	0.08 :	10.6 :	4.6 :	0.002 :
Jupiter	5.2	0.048	1.3	11.86	318
Saturn	9.5	0.056	2.5	29.46	95
Uranus	19.2	0.047	0.8	84.07	14.6
Neptune	30.1	0.009	1.8	164.81	17.2
Pluto	39.5	0.250	17.2	248.53	0.0025

removed through gravitational action of Neptune's large satellite Triton, the largest most massive satellite in the solar system. Triton has a diameter of 6000 km, and our Moon only 3476 km. Both, Moon and Triton, are larger and more massive than Pluto.

Note that the collective names for the groups of planets derive from the most massive planets in each group. Of the Earth-planets only, Earth and Mars are within the ecosphere – the region where the temperature permits life, as we know, though Mars is a bit cold. (Venus definately is too hot!) Note also that the Jupiter-planets appear to fall into two 'pairs', Jupiter and Saturn, and next Uranus and Neptune, which have small masses compared with Jupiter and Saturn.

We note the following *geometric properties* of planetary orbits (excepting the asteroids and comets).

(1) *The planetary orbits are nearly circular.* Exceptions appear to be Mercury, Mars, and Pluto. Apart from Pluto's abnormal situation we are dealing here with orbits of the objects of relatively small mass. The orbits of the heavy terrestrial planets, Venus and Earth have strikingly small eccentricities. The orbits of the Jupiter-like planets also have comparatively small eccentricities.

(2) *The planetary orbits are nearly coplanar.* Exceptions are Mercury and Pluto (inclinations of 7° and 17°), the planets nearest and furthest from the Sun. However the extreme distances from the average ('invariable') plane, as time goes by, the planets remain within 0.05 AU from the general plane. In other words the planetary orbits fit within a disk only 0.1 AU thick, of large 'equatorial' extent. The most distant planet Pluto requires an equatorial diameter of 80 AU for this disk, a matter of importance in explaining the origin of the planetary system.

(3) *The planetary orbits are co-revolving in the same general direction, obeying Kepler'laws (Chapter 1).* The same is also true for the majority of satellites and of the axial rotation of all planets (except Venus, Uranus, and Pluto).

(4) *The famous Titius–Bode 'law'.* The relative distances of the planets from the Sun are approximately expressed by $(0.4 + n \times 0.3)$ AU, where

$n =$ 0 for Mercury
       1 for Venus
       2 for Earth
       2^2 for Mars
       $(2^3$ for Asteroids)
       2^4 for Jupiter
       2^5 for Saturn
       2^6 for Uranus
       $(2^7$ for Neptune, very approximatively).

Otherwise expressed: With the exclusion of Pluto, and placing the average asteroid at 2.8 AU, each subsequent orbit, starting with Venus, is about 1.8 times as large as the preceding one. While this law has often been considered just a numerical coincidence, it may very well be related to the origin (and evolution) of a large number of companions to a massive primary. See Chapter 8.

The same 'law' is revealed within the satellite systems of Jupiter, Saturn, and Uranus. This may be no accident but may contain important information about the origin both of the solar system and of the satellite systems of several planets.

C. Physical and Chemical Properties of Planetary System

The *Earth-planets* are small and dense, ranging from 3.95 for Mars to 5.50 for the Earth, suggesting heavy material, *rocks* and *metals* such as iron and nickle. They rotate slowly, the Earth at one day being fastest. Their surface temperatures are largely above the freezing point of water, their maxima ranging from about 300 °C for Mercury and Venus to 25 °C for Mars. They have little hydrogen except in the form of surface water (or ice) for Earth and Mars.

The most important gases in the atmospheres of Earth and Mars are molecular nitrogen (N_2) and oxygen (O_2), carbondioxide (CO_2), and water (H_2O). The atmosphere of Venus contains CO_2 and N_2, that of Mercury atomic hydrogen (H_1) and molecular nitrogen (N_2). Mercury and Venus have no known satellites, the Earth has one, Mars has two.

The *Jupiter-planets* are large and have low densities, ranging from 0.69 for Saturn to 1.8 for Neptune. They rotate rapidly in periods ranging from less than 10 hr for Jupiter; to perhaps 16 hr for Neptune. The resulting centrifugal effects are evident in the form of the obvious oblateness of the Jupiter-planets. Their surface temperatures are well below $-100°$ C (below the freezing point of water). Their atmospheres contain methane

(CH_4), and ammonia (NH_3); molecular hydrogen (H_2) is found in the atmospheres of Jupiter and Saturn.

They have many satellites, the last counts ranging from 2 for Neptune, to as many as 21 for Saturn. Most of Jupiter's satellites are only a few hundred kilometers in diameter, and most likely are captured asteroids. The four Galilean satellites (Io, Europe, Callisto, and Ganymede, with diameters 3640, 3100, 5000, and 5270 km) are much larger, comparable in size with our Moon (3476 km) and with Mercury (4880 km). These four satellites physically qualify as planets. The same holds for the larger satellites Titan (5800 km) of Saturn, and Triton (6000 km) of Neptune.

Jupiter has Sun-like properties in the sense that it radiates more energy than it receives from the Sun. Must we consider Jupiter as a miniature Sun, with a planetary system? The discovery by Galileo in 1610 of the four bright satellites at that time certainly lent strong support to the Copernican view of the Sun's planetary system. The two groups of planets are separated by the belt of asteroids, whose orbital radii range from about 2.2 to 3.3 AU.

The largest four asteroids are Ceres (diameter 965 km), Pallas (560 km), Vesta (390 km), and Juno (240 km). Generally asteroids of which some 3000 have been named, have been discovered so far are less than a few kilometers in diameter.

Are the ringstructures of Jupiter and the other Jupiter-planets mini-analogies of the asteroids belt? Not certainly, since these rings are entirely flat and the orbital planes of asteroids by no means are anywhere nearly coplanar.

Sun and stars consist primarily of hydrogen (70% by mass) and helium (28% by mass). Other elements, conspicuous on Earth such as carbon, nitrogen, or oxygen account for only 1.5% in Sun and stars.

Anticipating the next chapter on origins we may at this moment surmise that at an earlier time hydrogen and helium, and part of carbon, nitrogen and oxygen have escaped from Earth-planets. The remaining rock-like material such as silicates, oxides, and other elements must be the small residue of originally much heavier cosmic material.

The heavier but much less dense planets Jupiter and Saturn were able to retain a much higher proportion of any original hydrogen and helium; some 80% and 67% for Jupiter and Saturn respectively. These planets are quite similar to the Sun, with mean densities 1.3 and 0.7, respectively, as compared with the Sun's density of 1.4. Jupiter and Saturn are probably mostly gaseous and liquid, with only a small solid nucleus with a mass of some 20 to 30 times that of the Earth of heavier rock-like material.

The original primeval nebula presumably was 98% gaseous-hydrogen and helium which are still the principal components of Jupiter and Saturn. The primeval nebula was 1.56%: ice-like (H_2O, CO_2, CH_4, NH_3) now in Uranus, Neptune, and comets. The remainder, 0.44% is accounted for as rock-like (silicates, oxides of heavier elements), principally in the Earth, other Earth-planets and asteroids.

A survey of principal geophysical knowledge of the planets of our solar system is given below:

	Diameter	Density	Material
Mercury	4 880 km	5.4	rocks and iron
Venus	12 100 km	5.2	rocks and iron
Earth	12 750 km	5.5	rocks and iron
Mars	6 800 km	3.9	rocks and little iron
Jupiter	142 800 km	1.3	hydrogen
Saturn	120 000 km	0.7	hydrogen
Uranus	55 800 km	1.0	hydrogen components
Neptune	48 000 km	1.7	hydrogen components

It is apparent that there is the conspicious occurrence in *pairs* (two and two) of planets: Venus and Earth among the *Earth-planets*, Jupiter and Saturn among the *Jupiter-like planets*. This grouping in pairs may also exist in the systems of satellites, strikingly for the four Galilean satellites: Io, Europe, Ganymede, and Callisto, all primarily stone material (Erich Litzroth, DDR).

The somewhat smaller Uranus and Neptune are somehow 'between' the Earth-planets and the massive Jupiter and Saturn. They consist primarily of the 'ices': frozen CO_2, H_2O, NH_3, CH_4, and other compounds primarily of the carbon, nitrogen, and oxygen with hydrogen, not unlike the building material of comets.

D. Comets and Extent of Solar System

Comets consist of ice-like material in which small stones are imbedded. Their orbits are not organized in any systematic manner. In contrast with the planets, cometary orbits are not co-planar, not co-revolving and not circular. Highly eccentric orbits and near-parabolic orbits are common; initially hyperbolic orbits have not been established with certaintly i.e. comets are not likely to be visitors from interstellar space. An analysis by Oort has shown that comets form a huge cloud, containing some 100 000 000 000 ($10''$) comets, extending up to some 20 000–50 000 AU from the Sun. This cloud therefore has a diameter of one or two lightyears i.e. up to almost half way to the nearest star Alpha Centauri, which has a distance of 4.3 lightyears.

Their arrangement and state of motion finds its galactic equivalent, on a very grand scale, in the 'populations' of our Milky Way system; we may speak of a 'halo' of comets, as contrasted with the 'disk' of Earth- and even more the Jupiter-planets. The total mass of all these comets is not more than a few Earth masses.

A comet approaching the Sun in its orbit may develop a tail of volatile matter pushed away by solar light pressure. These tails have been found to contain CO_2, CH_4 and other components, suggesting similarity in materials with Uranus and Neptune.

CHAPTER 8

ORIGINS

We shall briefly review ideas about the origin of galaxies and stars, discuss the origin of the solar system and finally attempt to define the distinction between stars and planets.

A. Galaxies; Solar System

It is generally agreed that some 10 000 000 000 yr ago the part of the Universe known to us was contained in a very much smaller portion of space than it is now. All matter consisted of an exceedingly concentrated 'gaseous' mixture of the elemental particles: primarily protons, neutrons, and electrons, at a tremendously high temperature of thousands of millions of degrees. The explosion of this original arrangement of matter sometimes referred to as the 'Big Bang', is still observed in the general expansion of the Universe of galaxies. Within the first hour after the explosion the temperature dropped to a point where the protons and electrons began to combine into the nuclei of the chemical elements. Deuterium (heavy hydrogen) nuclei were formed from the combination of one proton and one neutron. Helium nuclei ($2H_4$) resulted from the merging of two protons and two neutrons – and so on. As these nuclei combined with additional protons or neutrons, nuclei of heavier elements were formed. The creation of all the elements, as they exist now in the Universe, presumably took place during the first half-hour of the original expansion. After this, hydrogen still remained the preponderant element in the Universe.

The primary explosion was accompanied by turbulent motions, i.e. irregular currents, which gradually molded the chaotic distribution of material into more regular forms by gravitation. Locally the mutual attraction of particles led to concentrations of matter, and such groupings of matter generally whirled into a state of rotation. The concentrations of matter combined gradually into larger formations which again were in a state of rotation. A large body formed in this way retains its rotation. As the internal turbulence diminishes, the rotating object becomes flattened. The rotational speeds increase toward the center; the resulting shearing effects lead to turbulent cloud formations which by the rotation are spooled up in the form of spirals or *proto-galaxies*, the forerunners of spiral galaxies as we now know them. Spiral formations by turbulence are common. Examples are whirlpools in a disturbed flow of water, curling flames in a woodfire, tornadoes in our atmosphere, sunspots in the Sun's atmosphere. Cream poured into a cup of coffee or tea frequently leads to spiral formations. These often have a remarkable resemblance to cosmic spirals such as the well-known 'whirlpool' nebula M51. In the distant future, the rotation of spirals will cease, because the energy of rotation will be dissipated, essentially by the friction of the turbulent motions, the end result may be globular systems of matter.

B. Stars, Double and Multiple Stars

We consider that stars (including our Sun) resulted from concentrations of gaseous nebulae in spiral arms of galaxies. Small dark globular clouds are observed, something like one lightyear in diameter; these 'globules' may be protostars. Gradual contractions of these *proto-stars* raised the internal temperatures. And if the protostar was sufficiently large, nuclear reactions would start, resulting in creating a 'live' star, whose energy output would be primarily determined by the mass of the star.

These primeval concentrations rotated slowly and during further contraction rotated faster and faster. The amount of angular momentum in the rotation and the internal motions of the nebula are several hundred times larger than the maximum momentum which a star can contain. This explains why practically all stars have one or more companions in which much angular momentum can be stored. The existence of double, triple up to sextuple stars and more is thus explained.

At least half of all stars are members of a binary or multiple system. Abt and Levy have found that nearly all solar-type stars have a visible or invisible companion. Apparently nature solves the problem of angular momentum by storing the excess angular momentum in the orbits of one or more companions, which can be either stars or planets. Since the problem of angular momentum is universal, occurring in the formation of each star, it seems evident that, at least, for stars which have no visible companion a planetary system must exist. Millions of stars (only 1% of all the stars in the Milky Way System) may therefore possess a planetary system. The discovery of two planet-like companions for Barnard's star (Chapter 15), appears to support this opinion.

A *bona-fide star* is astrophysically defined as a self-luminous celestial body, whose mass is sufficiently large and hence the internal temperatures sufficiently high, so that thermonuclear reactions are ignited. Theoretical considerations show that this is the case for masses larger than about 8.5% of the Sun's mass. Several stars approaching and even below this theoretical limit have been observed; a good example is Ross 614B, first discovered as an unseen object, later detected visually and photographically. This, the fainter component of the double star Ross 614AB, has a mass of only 0.062 times the Sun's mass.

The astrophysical definition of a star has nothing to do with *cosmogony*, i.e. origin and evolution, of a star or system of stars. There is a great abundance of double (binary) and multiple stars possibly resulting from protostars sufficiently close to each other, to form double or multiple systems, whose gravitational stability would be insured for a long time. Double star orbits are elliptical with eccentricities ranging from 0 to 1.0. The mass ratios of the components range from 1 to 0.1.

Finsen and Worley's catalogue (1970) show that the observed eccentricities (in stable orbits) range from 0 to 1.0 (79 items). There are few eccentricities for e 0.2 and e 0.9; most are between 0.2 and 0.9 averaging 0.575, with a flat maximum from 0.4 to 0.8, within half of the total number occur. According to Ambartsumian (1937, *Astron. J. USSR* **14**, 207), theory gives an average eccentricity of 0.71. According to Harrington, theory gives an average of 0.68, more than half above $e = 0.60$ and hardly any very small or zero eccentricities (81 items) (1977, *Publ. Astron. Soc. Pacific* **89**, 400–401; 1976, *Astron. J.* **80**, 1081).

In the cosmogony of a double – (or multiple) star system the relative mass ratios are characteristic, the masses are comparatively large, 'comparable' in size. The mass-ratios are 2 : 3, 1 : 4, or something like that. Even mass ratios of 1 : 10 are still conceivable; the orbits are still elliptical, some of the companions may be invisible.

Anticipating Chapters 11ff we find that the orbits of astrometric binaries with one visible component are also strongly elliptical even when the mass of the companion is below the critical value of 0.10 $\mathcal{M}_\odot$. It is entirely possible that such a system with an invisible component, cosmogonically speaking is a double star, would therefore indeed be a 'star planet'.

An exception is the one extra solar planetary system for which nearly circular orbits are indicated (Chapter 15).

C. Origin and Evolution of Solar System; Disk

There is no theory of the origin and evolution of the solar system that is satisfactory from all points of view. We shall, nevertheless, outline some attempts generally agreed upon by astronomers and first foreseen by Immanuel Kant (1724–1804). These ideas should be valid also for planetary systems of stars other than the Sun. Many of the thoughts and ideas relevant to the cosmogony of the present topic are found in Su-Su Huang's articles with the pregnant titles 'Planetary Systems and Stellar Multiplicity' (1977), *Revista Mexicana de Astronomia y Astrophysica* **3**, 175–187), and *Extrasolar Planetary Systems* (1972), which have been incorporated in this section. I have also benefited from discussions with Hans-Jürgen Treder. Important information has been contributed by recent 'space' research.

The striking geometric and kinematic properties of the solar planetary systems have been presented in the preceding Chapter 7. We recall the near coplanarity, near circularity and corevolving of the planetary orbits, also for the majority of satellites as well as that of axial rotation (excepting Venus). Then there is the approximate geometric sequence of sizes of orbits (law of Titius–Bode).

These regularities are not direct consequences of a fundamental law of nature such as the law of gravitation, which permits other properties; they must be the result of the *manner* in which the solar system came into existence and therefore give specific indications of the origin of the solar system. The difference between the terrestrial and the Jovian planets and the halo of comets gives important information about the origin of the solar system. The comet cloud shows that our solar system extends to half the distance of the nearest stars; the same must hold for other stars. This is a strong indication that the solar system must have condensed from a gas cloud measuring at least several lightyears.

The average density of the system distributed over a cube with a side of say six lightyears (the average distance between stars in our neighborhood) results in a density of about 10 hydrogen atoms per cm^3; this is close to the density of gas clouds in the

Milky Way, such as the Orion nebula, in which at present one witnesses the *formation* of stars.

Moreover the comet cloud indicates that at the formation of the solar system, several, 10^{11} small ice-like 'cometisimals', or 'comet pieces', were formed. Along the orbits of old comets one finds countless small stones, meteorites mostly only a few centimeters in size now separated from the comet, which sometimes penetrate the Earth's atmosphere as 'meteors' of '*falling stars*'. These together with an analysis of the gases in comet tails show that comets consist of closely packed ices and stones. The chemical composition of comets reminds of the outer planets Uranus, Neptune, and to a lesser degree of Saturn. The rings of Saturn (and probably other Jovian planets) consist of small stones covered with a thin layer of ice. Current opinion is that the outer four, Jovian planets originated from the combining of many millions comet-like objects and that the now remaining comet cloud is only a slight remnant of a much denser and heavier primeval cloud, which must have existed during the formation of the planetary system.

The first physical theory about the origin of the solar system or rather the planetary system of the Sun is the so-called primeval nebular hypothesis of Immanuel Kant and P. S. Laplace (1749–1827). Apart from the coplanetary and corevolving properties they were impressed by the disproportionate amount of angular momentum (Σmvr) ('spin'), contained in the planets compared with the Sun. Of the total angular momentum of the solar system 99.5% is in the planets, 60% for Jupiter, – mostly in their orbital motion –; their total mass is only 0.14% of the total mass of the solar system. On the other hand the Sun, with 99.86% of the mass of the system has only 0.5% of its angular momentum. The total angular momentum is small compared with the total angular momentum of a double star.

Angular momentum cannot get lost. Kant and Laplace proposed that the solar system originated from a tenuous primeval nebula, which rotated very slowly and condensed due to the mutual gravitation between particles in the nebula. As a result and because of the law of conservation of angular momentum, the nebula started rotating faster and became flattened. At a certain moment the socalled centrifugal effect became so strong, that in the 'equatorial' plane of the nebula, successively a number of rapidly rotating rings came into existence. These were not stable and broke into pieces, from which the planets and their satellites condensed. After thus losing much angular momentum the remaining central portion of the nebula contracted and formed the Sun.

Later calculations have abandoned the ring hypothesis. But it is generally accepted that the solar system condensed from a *contracting rotating primeval nebula*. This nebula perhaps one lightyear in diameter, is assumed to have a slight initial rotation, with is more likely than no rotation at all, considering the general state of turbulence, an inherent cosmic condition. As the cloud contracts under the influence of gravitation, the rate of rotation increases, since the total angular momentum must be preserved. Almost the entire mass is concentrated in the Sun. The rotation of the Sun slows down through the escape of material from the outside which is rotating too fast. The gaseous material and dust surrounding the Sun or central star form a rotating envelope or protoplanetary *disk*; this implies a transfer of angular momentum from the Sun or central star to the surrounding medium. The planets there are formed from concentrations in the form of 'eddies', possibly through intermediary 'rings'. Nearly the total angular momentum is the orbital momentum of the planets. The total angular momentum however is small compared with the total angular momentum of a binary star.

A *disk* is one of the most *prevalent shapes in the Universe* because of its easy formation, as soon as the medium surrounding Sun (star) has acquired sufficient angular momentum. Cosmic disks are to be expected and are found. Examples are the observed rings of the Jupiter-planets. These are also inferred rings around Epsilon Aurigae and other stars but with strikingly large masses. It is expected generally that the ratio's of the masses of the planetary companions to the mass of the central star amount to at most a few percent and generally would be much less of the order of 0.1%. These planetary companions in agreement with their cosmogony from the accretion for a primary 'starnebula' must then move in nearly circular orbits, with radii of the order of some astronomical units. Generally there should be several planets, with different masses, and their orbits should be nearly coplanar. A striking example of this are of course the Jupiter planets in our solar system (Table 7.1). As to possible planetary systems beyond our solar system, we shall return to this subject in Chapter 15.

A planetary system thus evolves from a rotating disk of gaseous and dust particles, that comes into being after the star has already been there and therefore is a residue left over from the process of formation of the central star.

D. Binary and Multiple Systems

Binary and mutiple systems be formed in this way because they do not show the characteristics of having come out of a rotating disk. However, a planetary system could evolve around a binary component, provided the dimensions of such a system would be sufficiently small to insure long-range gravitational stability (Harrington, *Astron. J.* **80**, 1081, 1975).

In considering the possible existence and discovery of extrasolar planets, Jupiter-planets appear to have a good chance to be revealed by existing techniques, while the chances for discovering Earth-planets (mass 0.003% of Sun) seem to be very remote indeed.

Planetary formation on the other hand, depends on the size, turbulence, density, composition, and temperature of the flattened protoplanetary disk – around a new star. The size distribution of the resulting planets probably differs from the size distribution of the low-mass stars; their compositions are probably very similar.

For stars of sufficiently large mass, low-mass companion objects similar to the giant planets in our solar system could result. Astrometric perturbations have revealed unseen companions with a range of below one tenth solar mass to as low as Jupiter or Saturn masses (Chapter 13).

The material thrown off by the Sun consists, for the greater part, of hydrogen and helium, the two elements which are predominant in the Sun. Only a very small portion is formed by the more massive atoms, which gradually combine into solid dust particles of material common on Earth, such as iron-oxides, silicium compounds, ice crystals, and water. The planets are formed from a gradual accretion of these dust particles. Collisions between the particles lead to a more or less geometric spacing of nearly circular orbits for the planets in the sense that each successive planet is about twice as far from the Sun as the preceding one. The observed ratios are not too far below two (if we include the 'average' asteroid). There is the same agreement between theory and observation for the satellite systems indicating that the latter were formed in a similar manner.

The law of Titius–Bode (Chapter 7) has been supported by computer-analysis (S. H. Dole, *Icarus* **13**, 494, 1970). This law is the approximate geometric relation of the sizes of the planetary orbits, if one includes the average asteroid. Each planet is somewhat less than twice as distant from the Sun as its predecessor. Starting with a disk consisting of gas and dust, the orbits of a number of orbiting *planetesimals* become more and more circular. Each planetesimal grows by sweeping up dust and gas from the nebula. Calculations show, that planetisimals as a result of collisions will combine into larger or smaller orbits.

A planetesimal may be defined as a small body into which the primeval nebula condensed and from which the planets formed. Within only a few million years the large planetesimals will sweep up the smaller ones. The ultimate result of the calculations is always a limited number of large planets at regular distances from the Sun. These results signify that the distances are such that the orbits do not disturb each other. In particular the asteroids at an average distance of 2.8 AU from the Sun are not perturbed. Asteroids at other distances were perturbed, and may have been responsible for the bombardements of the terrestrial planets during the first half milliard years. That no complete planet was formed is ascribed to the proximity of the heavy planet Jupiter, which already had thrown the greater portion of the asteroids from their orbits, before they had the time to combine into a planet. The remaining asteroids at about 2.8 AU from the Sun, were too small in number and mass to form a planet.

E. Stars and Planets

Even if we limit ourselves to stars like our Sun, i.e. stars equal or comparable to the Sun in mass, surface temperature, luminosity i.e. a 'yellow' Main-Sequence star, there would appear to be millions of possible candidates for the development of planetary systems. There is reason therefore, to consider seriously the possibility of the existence of planets, stellar 'chips', near stars, other than our Sun and a search for such objects appears to be a valid astronomical objective. Pending striking advances in direct observational techniques, indirect gravitational methods exist and were initiated as long as half a century ago. We shall go in detail about the methods and programs dealing with the search for extra-solar planets, even if they cannot be 'seen' at the present time.

Remains to be considered how we define and distinguish planets. Theory has to come to the rescue and helps us by pointing to *mass* as the criterion. A *bona-fide* star should have a mass of at least 0.085 of the Sun's mass in order to be able to derive its energy from conventional nuclear reactions. However visible stars with mass smaller than this critical limit have been observed (Chapter 6). Theory also indicates that an object more massive then our planet Jupiter, will have some internal nuclear reactions, and it would seem therefore that a mass of about 0.1% of the Sun's mass is the critical upper limit for a 'cold' planet. This leaves a gap between 0.1% and 8.5% of the solar mass. There is no reason why such objects could not exist, even though they would not radiate (perceptibly) i.e. would be dark, invisible and potentially unseen unless they would be close enough to a *parent* star to shine by detectable reflected light or cause observable perturbations.

There is observational evidence for such objects, which are sometimes called *black dwarfs, brown dwarfs, star-planets,* or *sub-stellar objects.* The latter designation also would include planets as we know them in our solar system. Half a dozen sub-stellar objects, including two extra-solar planets, all companions to visible stars, have been detected by the gravitational approach.

An interesting designation, for classification purposes is the word *star-planet* used by Harlow Shapley (1885–1972) in 'Life Among the Dwarfs' in his inspiring book *Beyond the Observatory,* 65–66, 1967 (Charles Scribner's Sons, New York). We adopt the following guidelines:

Stars	Masses larger than 8.5% of Sun
Star-planets	masses between 0.1% and 8.5% of Sun
Planets	masses up to 0.1% of Sun

Our tentative knowledge of planets and other invisible objects of small mass beyond our solar system is essentially mostly limited to information about *orbits* of these unseen objects, i.e. orbital dimensions, eccentricities and in one case, the relative orbital inclination of two orbits. This kind of information aided by theoretical and cosmogenic considerations, is all we have, and we must use it analyzing and understanding our limited knowledge of unseen companions, star-planets and planets in our cosmic neighborhood.

To distinguish between stars, star-planets and planets we must not forget that our initial knowledge of planets is limited to our solar system. In searching for and identifying planets beyond our solar system we are to be guided therefore by what we know about the planets in our solar system. For stars our knowledge is far more extensive, and we need not consider our Sun as an example of what a star should be, though, as it turns out, the Sun is a well behaved typical star. We know of the existence of stars more luminous and many which are intrinsically much fainter than the Sun. And we have considerable information about stellar masses, the smallest measured values being about six percent of the Sun's mass. We also know that the majority of stars appear in groups of two or more, i.e. binary and multiple stars are the rule, and a single star like the Sun may well be the exception.

Apart from the obvious disparity in mass, there are the dynamical differences between stars, star-planets, and planets as exhibited in their orbital behaviour. The orbits of the planets in our solar system are nearly circular, coplanar and corevolving. Table 7.1 gives the orbital eccentricities and inclination relative to the plane of the Earth's orbit for the Jupiter-like planets and for the Earth-planets. Note the consistently small eccentricities and the near coplanarity for the four Jupiter-planets. Compare this with the orbits of the components of double stars where eccentricities range from zero to almost one. The orbits of star-planets behave as those of visual (and spectroscopic) binaries. In summary:

	Planetary systems	Binary star systems
Orbits	nearly circular	$0 < e < 1$
Mass-ratios	$\sim 10^{-3}$	up to 1.0
Orbit size	limited range in AU	wide range in AU

There are *generic* differences between the groups. A characteristic jump between 'star-planets' and genuine planets (both not self-luminous) should reveal itself in the orbits; a star-planet (with relatively large mass-ratio relative to central star) moves in an eccentric orbit, a genuine planet (with smaller mass-ratio) moves in a near circular orbit. For planets, of the Jupiter type, the distances to the central body should always be of the order of a limited number of astronomical units.

CHAPTER 9

HISTORICAL I

A. The Stars Sirius and Procyon

The name that enters the picture is that of Bessel, one of the ablest and most progressive astronomers of the first part of the nineteenth century. As early as 1824 Bessel noticed irregularities in the motion of Uranus; he was convinced that the perturbed behavior must be ascribed to the attraction of a planet, as yet unknown and unseen, beyond the orbit of Uranus. Bessel did not complete this investigation; but he did carry to conclusion a similar problem, leading to the first discovery of unseen stars by gravitational means. He studied the positions and proper motions of two of the brightest stars, Sirius and Procyon. Sirius is the brightest star in the sky, 8.6 lightyears away and, excluding the Sun, is the fifth nearest star. Procyon, also a star of the first magnitude, is 11.2 lightyears away; at present there are only 14 stars known to be nearer than this. As of many other stars, the positions of Sirius and Procyon on the celestial sphere had been regularly observed since the middle of the eighteenth century. From a study of these observations, Bessel suspected as early as 1834, that the proper motions of Sirius and Procyon were not uniform and rectilinear.

Uniform rectilinear motion, i.e., the law of inertia, is the rule for a single star sufficiently far away from other stars (Chapter 2). In the case of a deviation from this behavior, the qualifying condition obviously is not satisfied and the presence of a relatively close object is hypothesized to explain the observed irregularities. The center of mass of primary star and unseen companion follows a uniform rectilinear path, while both the visible star and its unseen companion each describe an orbit around the center of mass. Bessel noticed that the perturbations of Sirius and Procyon were periodic, i.e., went through repeated cycles, and on this basis he announced in 1844 the discovery of unseen companions both of Sirius and of Procyon. That these companion objects could not be seen did not deter Bessel from his interpretation. He wrote: But light is no real property of mass. The existence of numberless visible stars can prove nothing against the existence of numberless invisible ones.

The implications of his discoveries were well realized by Bessel:

For even if a change of motion can, up to the present time, be proved in only two cases, yet will all other cases be rendered thereby liable to suspicion, and it will be equally difficult, by observations, to free other proper motions from the suspicion of change, and to get such knowledge of the change as to admit of its amount being calculated.

Bessel's analyses were tested by others and confirmed although his conclusions were not verified until later. Nearly two decades later (1862) the companion of Sirius was first seen by A. G. Clark, with the 46 cm refractor now at the Dearborn Observatory. The very faint companion of Procyon was not seen until 1896 by J. M. Schaeberle with the

91 cm refractor of the Lick Observatory. The orbital characteristics of both, Sirius and Procyon, are now well known. Sirius and its companion revolve around each other in 50 years; for Procyon the period of revolution is 40.6 years. The primary of Sirius is 26 times as bright as the Sun; the companion has only 0.008 times the Sun's luminosity; the masses of Sirius and its companion are 2.1 and 1.1 times the Sun's mass. The primary of Procyon is seven times as luminous as the Sun, while the companion has only 0.0005 times the Sun's luminosity; the masses are 1.7 and 0.6 times that of the Sun. Although both companions are intrinsically of low luminosity, their large masses insure their classification as stars (Chapter 4); they are white dwarfs or degenerate stars.

B. The Planets Neptune and Pluto

The bright, naked-eye planets had set the limits of the solar system at the orbit of Saturn. In 1781 a more distant planet had been discovered visually by Friedrich Wilhelm Herschel. This object is barely visible to the naked eye; in small telescopes it looks like a star. It was not recognized as a planet until Herschel happened to note its change of position from night to night on the background of stars. Herschel recognized that the new object – later named Uranus – described an orbit under the gravitational influence of the Sun and thus was classified as a planet.

After some fifty years of observation it was found that the path of Uranus was not in complete harmony with the path expected after allowance was made for the known perturbatins, due principally to Jupiter and Saturn. In 1845 the discrepancy between theory and observation amounted to more than two minutes of arc, truly an intolerable amount for astronomers. This led to the first discovery of an additional, previously unseen planet in our solar system based on perturbations in the orbit of Uranus. To find this new planet would have been a natural sequel for Bessel after his discovery of the two unseen stars; but Bessel died in 1846. Meanwhile, the problem had been attacked by the mathematician John Couch Adams (1819–1892) of Cambridge, England. He computed the orbit of the unseen planet responsible for the perturbations of Uranus and submitted his results to the Greenwich Observatory; unfortunately, the telescopic check for the object was delayed. The same problem had also been investigated by the French mathematical astronomer Urbain Jean Joseph Leverrier (1811–1877). At this request, John Gottfried Galle (1812–1910) of the Berlin Observatory turned his telescope on September 23, 1846, in the direction calculated by Leverrier and with little delay identified the new planet, about one degree of arc from the calculated position, and easily identified through its disk of about three seconds of arc in diameter. The discovery of the new planet, later named Neptune, was another triumph for the reliability of the law of gravitation and its power as a tool for detecting the existence of unseen objects.

In this chapter I have chosen to report the *Discovery of the planet Neptune excerpted from 'De Sterrenhemel' by F. Kaiser (1808–1872), published in 1847.*

Kaiser was director of Leiden Observatory since 1837. He initiated a new observatory. He was one of the most careful observers of his time and inspiring teacher, both in the university and through his popular writings.

(Freely translated from Kaiser's leisurely Dutch):

Perturbations recently led to a discovery, so striking and so beautiful that in the realm of science, one could not point to the like of it. That the perturbations, which the planets in their orbits undergo, are known with a remarkable accuracy, can be shown by the striking agreement between their calculated and observed motions. A single planet, however, over the past twenty-five years has embarrassed the astronomers by seeming to demonstrate that something was still lacking from our knowledge of the motions of the planets. After the discovery of Uranus by the older Herschel, it appeared that it had already repeatedly been observed since the year 1690 but, noting neither motion, nor a disk, had regarded it as a small fixed star. Very shortly after the planetary nature of this object had become known, one also possessed positional observations on the sky which extended over more than its period of revolution. Therefore one was soon enabled to calculate its orbit with a high degree of precision. The observations of the planet Uranus became numerous after its planetary nature had been recognized and the accuracy of these observations increased with the perfecting of practical astronomy. During three decades the astronomer Alexis Bouvard (1767–1843), since 1793 at Paris Observatory, considered the time ripe, from the available observations to construct new tables for the motion of Uranus since 1821, from which the positions on the sky, past and present, could be easily calculated. To be able to succeed, he had to subject the perturbations of Uranus since 1821 due to the other planets to a new and rigorous examination. It appeared that the perturbations were insufficient to explain irregularities in the planets motion. When he published his table in 1821, Bouvard therefore declared that another force must act on Uranus, besides the attraction of the known planets. Different explanations for this still unknown force were suggested, but those who had seriously studied the motions of the other planets, felt convinced that the perturbations could result only from the attraction of an as yet undiscovered planet, beyond the orbit of Uranus. After the year 1821 there was frequent discussion on this subject, but there the matter rested. It was well enough known how to compute from the locations and masses of a planet, the perturbation on another. But here the inverse problem had to be solved and the reverse of the, already difficult, problem, appeared to involve still greater difficulties. Moreover the unexplained perturbations suffered by Uranus were so small that one might wellnigh despair at the possibility to derive from them with any precision the point on the sphere, where the suspected planet at a certain time would present itself. And even if one would succeed to some degree in determining the position, years might still go by, until the planet, in the midst of hundreds of small stars, appearing even in a small portion of the sky, could be found out.

Meanwhile the observations became more numerous and more precise, through which the difficulties of any investigation should diminish. And, although it was a big undertaking to prove the existence of the still undiscovered planet and to determine its location, it would seem that the required effort was not larger or more difficult than those, which in this century one has had for other gigantic enterprises, carried out in a successful manner.

In the present case one cannot be assumed of hitting one's mark and the possibility is the principal reason why it lasted such a long time, before some one took the matter in hands, something which also one had tried to elicit in vain through prize competitions. Finally after more than a year, Leverrier, who already had become famous through other outstanding accomplishments, came forward with a new investigation concerning the planet Uranus. Remarking that A. Bouvard had not calculated the perturbations, which Uranus underwent through the attraction of the known planets, with desirable care, he subjected these (perturbations) to an entirely new investigation. Especially, taking into account the most recent observations of the planet Uranus it become more certain than before, that the perturbation of the motion of this planet could not be explained completely by the attraction of the known planets. The upshot of these studies was announced on 10 November, 1845 in a treatise by Leverrier, submitted to the Academy of Sciences in Paris. It gave a very precise determination of the amount of the still unexplained perturbation, but left its cause still entirely undecided. In a treatise of 1 June, 1846 Leverrier gave the proof that the perturbation was caused by the attraction of a planet which moved around the Sun, beyond the orbit of Uranus; he reported the location on the sky, around which the still undiscovered planet should present itself. On 31 August, 1846 Leverrier gave an approximate determination of the mass of the still undiscovered planet, together with the size and shape of its orbit, the one as well as the others, solely derived from the motion of Uranus. Finally, on 5 October, 1846 Leverrier communicated the last portion of his investigations, concerning the situation of the orbit of the new planet.

There wasn't exactly any hurry to seek out Leverrier's planet on the sky; only after Johann Gottfried Galle (1812–1910) in Berlin, favored by an extremely fortunate bit of luck, on 23 September, 1846 had found it at the location pointed out by Leverrier, the latters labours were valued at their true worth, and he was celebrated and praised in the same measure as with suspicion one had neglected him earlier.

The discovery of the planet by Leverrier is perhaps the most striking and most beautiful of numerous discoveries, of which astronomy can be proud, and may she, in the eyes of astronomers, not be the greatest and most important, her result is understandable to everyone. She has shown, to any one comprehensible evidence of the admirable heights to which astronomy has risen by now, and for centuries she will be adduced as a proof for the excellence of astronomy.

These circumstances make it all the more unjust to fail to appreciate the merits of another youthful mathematician, Adams, who had preceded Leverrier and had completed his investigation, when the latter (Leverrier) had barely begun. Already in September, 1845 Adams had proven from the motion of Uranus the existence of the not yet discovered planet, and had determined the point in the sky, where he would be found. However, extremely scrupulous, he submitted his investigation to the judgement of two of the most famous astronomers of England, Messrs. George Biddel Airy (1801–1892) and James Challis (1803–1882), who succeeded Airy as director of Cambridge Observatory in 1836. The former attached little credence to the surprising result obtained by Adams, without carefully testing its worth, the latter was too much absorbed in his own pursuits; neither one bothered particularly about the work of Adams. Finally on 29 July, 1846, after Leverrier had already announced the results of his calculations, Challis commenced observations to trace the planet. He continued these on 4 August and 12 August, but, occupied with work, evoked by the, at this time, numerous comets, he initially left his observations unreduced, so that they did not lead to anythink. When finally the planet had been seen by Galle. Challis, who was not favored by the luck which had saved Galle nearly all trouble, turned to the reduction of his observations. It appeared that already on 4 and 12 August among a multitude of stars he had determined the location of the new planet, which among his observations he had not been able to distinguish from the stars. The truth of all this is so perfectly proven that only blind prejudice could doubt it. However, the planet first received the name Leverrier, and changed later to Neptune at the proposal of D. F. J. Arago. Leverrier was idolized as a miracle of this century, and from many sides was rewarded in a generous manner. On the contrary Adams was repaid by some of his compatriots with doubt at his honesty, and from the side of the French with diatribes and caricatures. However, injustice can not change the truth. Leverrier was the first to make known the results, Leverrier had given continual explanations concerning the course of his investigations which compelled respect for his diligence and his talents. Adams with the help of others, wished to be introduced in the scientific world; he kept silent and saw how the palm of honor was snatched from him. May the work of Adams be rather less beautiful than that of Leverrier, which cannot yet be judged completely, concerning the principal point, the location where the object had to be sought, he yet arrived at the same results. Had he, at his own initiative just as Leverrier, announced his investigations immediately after their completion, had only one astronomer made a serious attempt to trace the planet, the latter would have been discovered before Leverrier gave his first treatise concerning the planet Uranus, and nobody would have thought of associating the name Leverrier with the newly discovered planet. The circumstance that one had already spoken for years in different works about the planet whose existence was presumed, the circumstance that two astronomers, almost simultaneously, entirely independently from each other, and starting with partly different observations, determined the point in the sky where this planet should show itself, proves to us, that the greatest honor of the discovery belongs to the high excellence of present day astronomy. It enabled both, Leverrier and Adams, to carry out their investigations, which they would not have achieved without the labors of their predecessors, without the availability of numerous, very precise observations.

After some time it became clear that even allowance for the gravitational attraction by Neptune did not fully permit an accurate prediction of the future path of Uranus. The possibility that there was still another unseen planet beyond Neptune presented itself. Leading among the investigators of this problem was Percival Lowell (1855–1916), who in 1906 at the Lowell Observatory in Flagstaff, Arizona, initiated a photographic search for the new planet whose existence he had inferred from calculations. An intensive systematic search was made and rewarded by the discovery of the new object on February 18, 1930 by Clyde William Tombaugh (1906), after an inspection of some two million photographic images. The new planet was finally located through its slight displacement on two photographs taken six days apart. The new planet, named Pluto, is a faint object, visible only in large telescopes. (At the moment it does not seem likely that any further, 'trans-Plutonian' planets will be easily discovered.)

A detailed presentation on this discovery of Pluto was prepared for this freatise at my request. This grateful preparation by C. W. Tombaugh, a personal presentation is given in Chapter 10.

Other 'classical' pre-photographic discoveries and analyses of unseen companions were based on the perturbations in the orbital paths of the binaries Zeta Cancri C in 1888 (Seeliger, 1914) and Xi Ursae Maioris A by Nörlund (1905). Both discoveries were based on visual micrometer observations, and later were fully confirmed from photographic measurements (see Chapter 12). Neither companion has yet been seen.

CHAPTER 10

HISTORICAL II

A. 'The Predictions and Discovery of the Ninth Planet, and the Extensive Planet Search', by Clyde W. Tombaugh

Prior to the discovery of Uranus by Sir William Herschel on March 13, 1781, little thought was entertained on the possible existence of planets beyond the orbit of Saturn. This is reflected in Herschel's interpretation of the object after he found it, thinking it was a tailless comet. Not until several months later was the object proclaimed to be the seventh planet.

Several decades later, the positions of Uranus were found to be deviating from its orbital parameters. After 1830, a few atronomers began to suspect that the deviation of Uranus was caused by the gravitational perturbation of an unseen planet beyond Uranus, accumulating to the intolerable amount of two arc minutes. This paved the way for the dramatic mathematical prediction of Neptune's position in the sky by Adams and Leverrier, of England and France, respectively, and its observational detection by Galle in Berlin in September, 1846. Both Adams and Leverrier pleaded in vain with observers in their respective countries to search for the predicted planet. It has always been a mystery to me why they did not borrow a spyglass and search for the planet themselves. Plotting a hundred stars freehand on a piece of paper and checking a few nights later would have been sufficient to have detected Neptune by its change in position.

The first serious search for a planet beyond the orbit of Neptune was probably Todd, using the 26-inch refractor at the United States Naval Observatory in Washington. He had in mind a planet at a mean distance of fifty-two astronomical units, with a diameter of 80 000 km, subtending an angle of 2 arc sec. Instead of depending on motion, he relied on a perceptible disk for its detection, using high manifying powers of 400 to 600. He searched along the Invariable Plane to a distance of one degree to either side. On thirty clear, moonless nights, between November 3, 1877 and March 5, 1878, he searched a strip from longitude 146.8 to 186.1 degrees in the constellations Leo and Virgo, (far from the star-rich fields of the Milky Way). He inspected 3000 stars down to the 13th magnitude for signs of a perceptible disk. He suspected many objects, which were reobserved on the following nights and again several weeks later; but, they were just stars – no motion.

Attempts at prediction were made by Forbes and Pickering, on the basis of the effects of unseen planets on comets. Indeed, Pickering predicted half a dozen different planets of widely different magnitudes in different positions in the sky. Of particular interest, Pickering's investigations led to a search at the Mount Wilson Observatory in 1919, by Humason, who took four plates around the position of Pickering's 'Planet O', with a 10-inch astrographic telescope, without success. Later, in 1930, after the discovery of

Pluto, images of Pluto were found on some of those plates, when it was known precisely where to look. Pickering pleaded with observers with suitable instruments to look for his predicted planets. In the issues of *Popular Astronomy* in the 1920's, Pickering made many drastic revisions of his predicted planets, which evidently discouraged would-be searchers.

Percival Lowell became interested in the problem of a trans-Neptunian planet early in the 20th century. Indeed, he initiated photographic searches as early as 1905. From 1914 to 1916, Lowell and his assistants made a vigorous search for his predicted 'Planet X' with a 9-inch Brashear telescope, loaned by the Sproul Observatory. By this time, they had acquired a blink comparator from Zeiss to scrutinize the pairs of plates. Lowell died suddenly in November, 1916, without finding his Planet X. No further search work was done until the completion of the 13-inch Lawrence Lowell Telescope in 1929.

It was at this point that I came into the picture. In the arly Spring of 1928, I completed my third homemade telescope, a 9-inch Newtonian of 79 inches (210 cm s) focal length. After several weeks of frustrating polishing, I obtained a parabolic figure of excellent quality. It yielded sharp images when using magnifying powers as high as 400 diameters. In the Autumn of 1928, I made many careful drawings of Mars and Jupiter at the eyepiece, and sent them to the Lowell Observatory. The timing could not have been better. Unknown to me at the time, the Lowell Observatory was forced to look for an amateur to operate the new photographic telescope, because of limited funds. Promptly, I received a letter from V. M. Slipher, the Director, asking many questions about my schooling, health, etc. Although, I had only a high school training, taking all the physics and mathematics offered, I studied solid geometry and trigonometry on my own, also, all the astronomy books I could get my hands on. I had spent hundreds of hours observing objects with my telescoe (from the Moon and planets to globular star clusters and galaxies). In the last week in December 1928, a final letter came inviting me to join the staff to operate a new photographic telescope on a 3-month trial basis. In June 1928, our wheat crop was a total loss from a very severe hailstorm, so I hired out to a neighbor several miles away to run his combine during the wheat harvest.

On January 14, I boarded the Sante Fe train at Larned, Kansas for the 27-hour trip to Flagstaff in a chair car. I did'nt have enough money in my wallet for a return ticket. I was nervously apprehensive, wonder what lay ahead. Dr V. M. Slipher met me as soon as I walked into the Flagstaff depot, and took me up Mars Hill to the Lowell Observatory.

The next day, I was taken to the new 13-inch telescope, not yet quite finished. Only then, did I learn that the new telescope was to be used to find Lowell's predicted Planet X. I was quite excited at the prospect.

The 13-inch Cooke Triplet lens (in cell) arrived on February 11, from the Alvan Clark firm (made by C. A. R. Lundin) in Massachusetts. It was bolted on to the tube and for the next several weeks I assisted V. M. Slipher in making various tests to ascertain the optical properties and performance of the instrument. It was found that 14 by 17 inch plates could be used effectively if the plates were bent in the plate holder to conform to the 'Petzval Curvature' of the focal plane.

After making a few one-hour exposure plates under Slipher's direct supervision, he said to me: "I think you can handle it. You are now on your own," and he left the dome. Apparently, he was satisfied with my self-taught knowledge of astronomy and instruments. I was entrusted to select the guide stars along the Ecliptic for the proper spacing of the plates, and processing (developing) of the large plates. However, the 'blink' examination of the pairs of plates would be carried out by 'the more experienced members of the staff'.

The photographic planet search got underway in April 1929. Although the telescope had been well designed and built, I was experiencing several operating problems. At times there was severe 'pulsing' in the sidereal driving, which I finally solved by operating the telescope entirely on the west side of the polar axis, and adjusting the counter weights to keep constant pressure of the worm against the teeth of the large (4 foot diameter) wormwheel.

Some plates had double star images. I discovered that this occurred when the long exposure passed through a critical hour angle of 42 min west of the meridian. This was caused by a slight 'chug' in the declination axis. Tightening the declination axis collars did not work. The double image problem was solved by swinging the telescope westward to chug the axis before setting the telescope to begin the exposure.

The most nerve-racking was the loud, sudden bang when a plate would snap in two during some of the one-hour exposures. I complained to Slipher about it. He replied: "You have to expect a few plates to break when you have to bend them in the plateholders." My observing schedule was so heavy that I could ill-afford using valuable observing time to replace plate casualties. So I studied the problem and changed the 'bending' procedure. I never had a plate break since.

Among the first pairs of plates taken were in the Gemini region of the Zodiac, according to Slipher's instructions, because it was the region favored by Lowell's predicted position for his Planet X. I can never forget Slipher'sλ astonishment when sample counting indicated up to 400 000 star images per plate in the winter Milky Way. In fact, he seemed rather overwhelmed. I made the foolish remark that: "I was glad I did not have to wade through the starry mess."

Since C. O. Lampland was away at Princeton on a sort of sabbatical leave for the Spring semester, only the two Sliphers (V.M. and E.C.) were available to blink the plates. Alternating with each other, they spent about two full weeks blinking the three pairs of plates covering the span of Gemini. They had hoped for a 'quick find' of Lowell's Planet X. As I learned later from personal experience, their two weeks of blinking was about three times too fast to be thorough in the scutiny of the plates. Unknown at that time, the images of Pluto were recorded on the Delta Geminorum pair, but they missed detecting them. I remember the day when V. M. Slipher took the last pair off the Blink-Comparator. It was quite evident that he had suffered a very disappointing defeat. At this season, Gemini was not favorably placed for photographing, being low in the evening western sky, which was often plagued with light haze. Also, the geometry was all wrong in regard to providing parallactic shift that would indicate the object was beyond the orbit of Neptune. Anyway, the Sliphers were looking for an object two magnitudes brighter than Pluto, as predicted by Lowell.

I kept on taking plates, progressing eastward through the Zodiac as fast as moonless skies and weather would permit. The Sliphers did no more blinking and by the end of the June lunation, about 100 plates had accumulated, only six of them blinked. This was beginning to worry me, but being a junior member, it didn't seem appropriate for me to ask questions. In early June, C. O. Lampland had returned from Princeton, but he did not attempt any blinking either.

One day in latter June, just after I had finished a strenuous run of night work taking plates, V. M. Slipher came to my office and informed me that they wanted me to start blinking the pairs of plates. I shuddered at the prospect of the task. Had they secretly become unsure of the validity of Lowell's prediction?

After labouriously blinking two pairs, I was very upset. I encountered several dozen asteroids on each pair. How was I to distinguish an asteroid from the sought Planet X? Strangely, I was never cautioned against taking the plates at the apparent near stationary positions of the asteroids. A feeling of dread and hopelessness came over me. The task seemed futile. Since I had no college training to my credentials, I had to stick with it if I wanted to remain in scientific work. It was better than pitching hay back on the Kansas farm.

Then I remembered observing during my youth, the angular motions of the planets from east stationary westward through opposition to west stationary, then reversing to easterly motion again. Also, that Saturn (the most distant) had the least amount of apparent motion of the three planets that I had observed. Suddenly, I saw the solution to my asteroid problem.

During the July–August rainy season at Flagstaff, I studied this problem in a quantitative fashion from the positions of Uranus and Neptune in the *American Ephemeris and Nautical Almanac*. If I took the plates strictly in the 'opposition' region of the Zodiac, I could readily distinguish asteroids from Planet X. Moreover, I could almost instantly deduce the approximate distance of a planet suspect by the amount of shift in position by parallax produced by the daily orbital motion of the Earth.

My attitude changed from despair to hope. Realizing the tremendous potential of the 13-inch plates for making discoveries, I might find a smaller, unpredicted planet, or maybe two, if I made a thorough search of the entire Zodiac belt. The blinking would be a long, tedious job, but it might be worth it. Perhaps Planet X was in a different portion of the Zodiac than Lowell had predicted. If the result was negative, at least we would know the content of the outer parts of the solar system.

Little did the astronomical world know of the frustrations at Flagstaff that occurred during the Spring and Summer of 1929.

In August, I talked to Slipher about the supremely important strategy of taking the plates at or near the 'opposition' regions of the Zodiac. Also, that most of the pairs of plates I had taken in the Spring of 1929, would have to be done over. Moreover, most of the pairs had no third plate for confirmation of planet suspects. If one presses the search to the magnitude limit, several dozen false planet suspects are encountered on each pair.

Not until I had read William G. Hoyt's manuscript for his book *Planets X and Pluto* (in 1978–1979) did I learn that Lowell had recommended taking the plates at opposition, in a quote from the Lowell Observatory Archives. After the discovery of Pluto in 1930, the Pluto images were found on two plates taken with the 9-inch astrograph in 1915. The dates those plates were taken were March 19 and April 7, respectively. I was shocked at this lack of proper procedure. For Pluto's position in 1915, the plates should have been taken in December.

With the ending of the Flagstaff rainy season, I started taking plates in Aquarius and Pisces. Since the Sliphers failed to find the planet in their blinking of the Gemini region in May, it seemed to me that searching other regions would be just as good. My plate taking marched eastward 30 degrees each month to keep up with the opposition regions. I encountered Uranus, which had exactly the amount of daily shift in position westward (apparent retrograde) that I expected. During the 'light of the moon' periods, I was able to keep up with the blinking. In Taurus, the great increase in the number of stars slowed down the blinking.

In January, I rephotographed the three plate regions in Gemini. The two western regions contained approximately 400 000 star images per plate. I decided to postpone blinking them, and placed the Delta Geminorum plates on the Blink Comparator. These contained about 160 000 star images each.

I had blinked one quarter of the pair when suddenly I encountered the images of Pluto. 'That's it!' I exclaimed to myself. Almost instantly, I knew the object was far beyond the orbit of Neptune from the small amount of shift in the six-day interval. My intense excitement can hardly be described. This was about at 4:00 p.m. on 18 February, 1930. I had taken the discovery pair of plates on 23 and 29 January, 1930. In the six-day interval, Pluto had shifted $3\frac{1}{2}$ millimeters westward from the 23 January position. The scale of the plates was 122 arc sec per millimeter. After I had started the first plate of the region, the seeing became very unsteady and produced swollen images. This plate on 21 January was unblinkable with either of the other two.

I removed one of the discovery plates and placed the 21 January plate on the Blink-Comparator. Almost instantly, I found the Pluto image 1.2 millimeters to the east of the 23 January position, which was perfectly consistent with the 6-day shift. I then felt 100% sure I had a planet nearly 10 AU beyond the orbit of Neptune. I then informed V. M. Slipher and C. O. Lampland, and they came to the Blink-Comparator room to see the images. The atmosphere was tense with excitement. Slipher said: "Tell no one until we follow it for a few weeks, and we are ready to announce the discovery. Mr Tombaugh, rephotograph the region as soon as possible." There was one concern. Pluto's photographic magnitude was about $15\frac{1}{2}$, $2\frac{1}{2}$ magnitudes (10 times) fainter than Lowell had predicted. Perhaps it was a very dense body, with a very low reflective power? Nevertheless, the shift was in agreement with Lowell's assigned distance.

The night of 18 February was very cloudy. On the following night, I was able to get another plate of the region. After I got the plate processed and dried, I placed it on the Blink-Comparator against the 29 January plate. After three weeks of no record, I expected Pluto to be about 11 millimeters further west. In a few seconds of blinking, I found it exactly where it should be.

Slipher instructed me to make a 5 by 7 inch film contact from the plate of the immediate region. On 20 February, Slipher, Lampland, and I went to the 24-inch refractor. Setting the telescope to the proper position, Slipher started checking the field of stars from the film. There it was, slightly farther west, a very faint star-like object with no visible disk.

I continued to take plates of the region with the 13-inch telescope for the next several nights. Lampland then photographed Pluto with the 42-inch reflector on every possible night for many weeks. Comparing blue and yellow plates, he established that the color was yellowish, unlike Neptune.

Also, Lampland took several long exposures in an attempt to find a satellite and thereby determine the mass, but without success. Lampland devoted his time to measuring precise positions. E. C. Slipher started experimenting with a box having different sized holes and different levels of illumination placed two miles away and studied with high powers on the 24-inch refractor. V. M. Slipher prepared an Observation Circular. The secretary and I addressed hundreds of envelopes and tucked in the Circular, ready to take to the post office for destinations all over the world on the selected date of 13 March for the public announcement of the discovery. On the evening of 12 March, V. M. Slipher sent a terse telegram to the Lowell Observatory Trustee, Roger Lowell Putnam in Massachusetts (Percival Lowell's nephew) to be transmitted to Shapley the next day for the Harvard College Observatory distribution.

Both the astronomical world and the public went wild with excitement. Newspaper reporters and magazine article writers swarmed over Mars Hill. It was bedlam. Telegrams poured in, congratulations and demands for more positions of the planet.

The early computed orbits were wild. The short interval of two months was much too short an arc to determine the semi-major axis and eccentricity. Soon various astronomers were finding images of Pluto on old plates, as far back as 1908. These older positions provided a much longer arc of the orbit, and a much better determination of the orbital elements.

The new orbital elements were in surprising agreement with those predicted by Lowell. It appeared that Lowell's prediction was fulfilled. But, one nagging question persisted. How could this faint, small objects have enough mass to have produced the residual perturbations on Uranus and Neptune? Lively controversies developed and persisted for the next several decades.

In latter 1930, some apprehension arose that the real Planet X was yet to be discovered. Several astronomers pleaded with the Lowell Observatory to continue the planet search. They commented with, "You have the instrument, the technique, and the skill, what else is out there?" Slipher was impressed with the thorough manner in which I was conducting the planet search and while I still had the enthusiasm, instructed me to continue the search.

I scanned every star image into the 17th magnitude over the entire Zodiac belt. The regions of Scorpius and Sagittarius were particularly tedious and time consuming because each plate had one million star images each.

Since Pluto had an inclination of over 17 degrees, there might be other bodies with even higher inclination and far from the nodes, which would place them outside of the Zodiac belt. Over the following years, I photographed many parallel strips to each side of the Zodiac, until I had covered 70% of the entire sky. Adequate overlaps were always made for time differences for possible highly inclined orbits for objects as close in as Saturn's orbit. There were thousands of faint planet suspects, everyone checked with a third plate and found to be false. No more planets showed up.

Kowal's mini-planet, Chiron, has an inclination of only six degrees. At the time, I searched the Zodiac (1930), Chiron was far out in its orbit beyond Saturn, and about $\frac{1}{2}$ magnitude too faint to be recorded in the 13-inch telescope plates. Had I made the search of the Zodiac in the latter 1930's or 1940 when Chiron was near the perihelion portion of its orbit, I would have discovered it.

I would have continued the planet search to an even higher Ecliptic latitude, but World War II interrupted the search because I was in the military draft. Nevertheless, I had already thoroughly searched all regions to 40 degrees and more north of the Ecliptic. Also, I had searched all southern hemisphere regions north of south declination 50 degrees, even Canopus and Omega Centauri. Hence, I would assert that Lowell's Planet X does not exist, since Pluto's mass was found to be only 1/400 that of the Earth. Over 14 years, I spent 7000 hr in actual blinking, and saw individually 90 million star images.

All of this vast search is described in more detail in my book *Out of the Darkness*: *The Planet Pluto* (211 pages) co-authored with Patrick Moore, and published by Stackpole Books of Harrisburg, Pennsylvania, in September 1980, and January 1981.

B. 'Planet X?' by Robert S. Harrington

Near the beginning of this century, both Percival Lowell and William Pickering carried out calculations to try to determine the orbit of a supposed planet out beyond the orbit of Pluto. The basis for this idea was the inability to accurately predict the motions of Uranus and Neptune, and it was the observed perturbations in the orbits of these planets that were used in their calcuations. On the basis of the calculations by Lowell, systematic searches were carried out for this planet, and the third one turned up Pluto. At the time it was believed that Pluto was the predicted planet, even though it was much fainter than would have been expected.

Over the 48 years after the discovery of Pluto in 1930, more and more data accumulated on the motions of the outer planets, which permitted more accurate estimates of the mass of Pluto. With each estimate, its mass got smaller and smaller, and there finally even appeared a tongue-in-check calculation that suggested the mass of Pluto would become imaginary in 1984 and the planet would disappear altogether. This raised increasing doubts over whether Pluto was indeed the missing planet, and the entire matter was settled for good in 1978 when a satellite of Pluto was discovered and its mass could be well determined. Pluto is a factor of 1000 too lightweight to have produced the effects observed in the orbits of Uranus and Neptune and therefore could not be Lowell's and Pickering's planet. Is their planet still awaiting discovery?

The accumulated new data on the motions of the outer planets has also become more accurate, and the problems plagueing Lowell and Pickering have not gone away. We still discover that, even though we use the best data and the best methods of computing orbits available, Uranus and Neptune still run off their predicted paths after a few years. However, the amounts of these run-offs are quite small, making it still unclear whether they are true planetary perturbations or systematic errors in the observations. We certainly can not rule out serious systematic observatinal errors in the old data from last century. Hence, there does not exist compelling evidence for yet another planet in the solar system. However, there is a lot of at least highly suggestive evidence, enough so that many astronomers are quite confident that there is something more out there and that the searches first suggested by Lowelel and Pickering should be continued.

There are two rather small observational searches under way at present. One is by Charles Kowal at Palomar Observatory, and the other is by Robert Harrington of the Naval Observatory. Both involve the same technique used by Tombaugh, which is to take photographic plates with wide field telescopes of selected areas of the sky. Two plates are taken, separated by a few days, and these plates are compared on a 'blink comparator' to try to locate any slow moving objects. The size of the motion indicates how far away the object is, and hence any planets would be easily spotted. Although both efforts have been underway since Pluto's satellite was discovered, both are very modest efforts covering small portions of the sky, and nothing has turned up so far.

What would this missing planet look like, anyway? The calculations to date suggest that it would be intermediate in mass between the Earth and Uranus. If it were a small gaseous planet like most of the other outer planets, it would be about the same brightness as Pluto and thus not too hard to find. The fact that it was not found by Tombaugh indicates it was in the part of the sky not surveyd by him (probably far South), which places some restrictions on where to look now. It would have on orbit with a mean radius of probably 50 to 80 AU's, which would put its period in the 400 year to 700 year range. There is reason to suspect its orbit, like that of Pluto, would have both inclination and eccentricity of significant size, and it may well be near aphelion now, making discovery somewhat more difficult.

An interesting, rather exotic idea has been suggested by Tom van Flandern and Harrington in connection with this missing planet. They suspect that it was initially between the orbits of Uranus and Neptune, and at that time Pluto was a satellite of Neptune (in terms of size and density, Pluto looks much more like a satellite than a planet now). The missing planet had a close encounter with Neptune back in the early days of the solar system, and this encounter: (a) dislodged Pluto from Neptune and put it in an orbit similar to its present one (the capture into resonant lock came later); (b) disrupted the rest of the satellite system of Neptune to the point that what is now Triton went into a retrograde orbit and Nereid went into a highly eccentric orbit; (c) the planet itself was thrown out into a much more distant orbit with high inclination and eccentricity. There is not much more to support this idea than some interesting numerical coincidences, but it is an intriguing idea.

And then what about 'Nemesis', the recently suggested distant stellar companion to the Sun (note, however, that this is not a new idea) that every 26 million years comes to perihelion on its very eccentric orbit? Perihelion is in the inner portion of the Oort cloud of comets, and the passage of the star stirs up the comets to the point where they rain down on the Earth as they swarm through the inner Solar System. This terrestrial bombardment raises massive amounts of dust into the atmosphere, to the point that the Sun's light is almost completely blocked out, bringing an end to all but the most primitive life forms, causing one of the mass extinctions of life that are now thought to occur periodically rather than randomly. Fortunately, this has no connection with the planetary perturbations of concern here. In the first place, the period of the star is many orders of magnitude greater than those observed in the planets. In the second place, the star is now thought to be at aphelion if it exists at all, and out there its tidal influences on the planetary part of the Solar System would be completley negligible.

CHAPTER 11

PERTURBATIONS

A. General

The Sun is the predominant factor in determining planetary motions. If the gravitational effect of other planets could be ignored, each planet-Sun binary would present the ideal example of the two-body problem; the planet would move around the Sun in an elliptical orbit subject to the law of areas; a case of perfect Keplerian motion would exist (Chapter 1). However, there are also gravitational attractions between each planet and all other planets. Since the net effect of these attractions is a small, but not negligible acceleration, the orbits of the planets are slightly distorted ellipses. These distortions are called *perturbations*, which the law of gravitation permits us to express in terms of the relative spacing and masses of the planets. It is one more proof of the reliability of the law of gravitation that paths, past and future, of the planets can thus be predicted more precisely than when the perturbations are neglected.

Two unseen stars, stellar companions of the bright nearby stars Sirius and of Procyon and two unseen planets in the solar system were discovered before the middle of the nineteenth century (Chapter 9), and later were seen.

We make allowance for the perturbation of the planets in so far as the latter are known to us. Suppose now that still other planets exist, not yet recognized. If such an unseen planet had sufficient mass, it would cause additional measurable perturbations in the paths of the known planets. Similarly, the presence of an unseen star or planet close to a visible star would be revealed in the course of time by the gravitational perturbation caused by the unseen object. Imagine what would happen if our Sun were invisible but still had the same mass. The Sun would dominate the planetary motions gravitationally, just as it does now. Kepler's laws of planetary motion would have pointed to the existence of the Sun, even if we would not have observed it directly.

Before entering on the promising and effective approach, i.e. the astrometric discovery and subsequent study of perturbations in stellar or planetary motions we hall first briefly mention other methods which permit the discovery of unseen companions to stars.

B. Spectroscopic, Photometric, and Eclipsing Companions

Periodic shifts in spectral lines of stars interpreted as changing Doppler effects (Chapter 2) indicate the presence of an unseen companion. Spectroscopic binaries are revealed through periodic variations in the radial velocity of an apparently single star. In the ideal case, two superimposed spectra are present, revealing the two components. Often, however, only one spectrum is observed; nevertheless, its periodic shift demonstrates deviations from uniform rectilinear motion, and hence the presence of an unseen companion. Spectrographic observations have been particularly suitable for stars which are apparently bright, regardless of their distance; they favour the discovery of relatively short-period binary stars because of their large range in velocity. None of these objects has ever been observed directly as a binary; even through the telescope they remain visually unresolved. The spectroscopic study of these close binaries has one disadvantage; it does not show the inclination of the orbit with respect to the line of sight. Hence, the information which can be gathered for individual binaries is limited, though statistically or in connection with other types of observations important conclusions about stellar properties have been derived.

As we have seen in Chapters 4 and 5, the majority of stars in our immediate neighborhood exists in groups of two (double or binary stars) or three (or more) (multiple stars). Early spectroscopic studies revealed a large number of stars, accompanied by invisible companions of substantial mass. As early as 1905 the spectroscopic observers, William Wallace Campbell (1862–1938) and Heber Doust Curtis (1872–1942), both pioneers in the field of spectroscopic binaries wrote:

In fact, the star which seems not to be attended by dark companions may be the rare exception. There is the further possibility that the stars, attended by massive (stellar) companions, rather than by small planets, are in the decided majority; suggesting, at least, that our solar system may prove to be an extreme type of system, rather than a common or average type.

We are tempted to conclude that stars lacking stellar companions seem to be the exception. As to companions of low mass, particularly planetary companions, the spectroscopic method is not (yet) effective and we have to rely on astrometric studies.

Faint companions of stars can be discovered if the orbit is seen on edge or nearly so. In such cases mutual eclipses of the two components take place periodically; the resulting variation in the observed brightness reveals the binary character of a star, which visually appears single. The best known example is the bright star Algol (Beta Persei) whose eclipses are repeated every 69 hr. Hundreds of *eclipsing binaries* (also called eclipsing variables) are known; aided by spectroscopic observations, they furnish important information about the masses, diameters, luminosities and densities of their component stars. Though planetary companions have not yet been discovered in this way, a planet may some day be found from the minute eclipsing effect on its star.

Remains the promising, at present most effective way, of discovering and studying unseen companions by *astrometric* means, from observed perturbations in the stellar (or planetary) motions. This method is best introduced by a review of the classical discoveries both of unseen planets in our solar system and of two unseen stars shortly before the middle of the nineteenth century (Chapter 9), to be followed by the discovery of yet another planet in the solar system (Chapter 10). These four unseen objects were later seen and photographed. Following these classical discoveries, two later perturbations, found from photographic studies of stellar paths, resulted in subsequent sightings (Chapter 14) of previously unseen companions. There are also discoveries of still unseen companions in visual binaries (Chapters 12 and 13).

We must be prepared for more complex perturbations, due to more than one companion (Chapter 15) and to the variable luminosity of either or both primary and companion(s).

C. Photographic Astrometry of Perturbations

From the earlier detailed orbital analysis given in *Stellar Paths* (Chapters 10–13), we summarize the relevant observations for the analysis of one perturbation. The remainders R from a solution for proper motion and parallax (and quadratic effect, if need be) may reveal a perturbation over the interval covered by series of observations. If the interval exceeds the period of the perturbation, the latter will be revealed and strengthened by a repetition of the pattern of the remainders.

Let us for the sake of illustration ascribe a perturbation to one companion with a constant luminosity from the primary. The *dynamical elements* period P, periastron T and eccentricity e may be derived from the time displacement curves R/t in any coordinate, ideally and preferably right ascension and declination. This method is of particular value for evaluating the eccentricity (and periastron passage) of an unresolved astrometric binary.

The geometric elements (B), (A), (G), and (F) refer to the *photocentric* orbit of an unresolved system and are given in parentheses to distinguish them from the geometric elements B, A, G, and F (natural elements or Thiele–Innes constants) of the relative orbit of primary and *one* companion. The geometric elements of the photocentric orbit are related to the Thiele–Innes constants as follows:

$$(B) = -\frac{\alpha}{a} B, \qquad (G) = -\frac{\alpha}{a} G,$$

$$(A) = -\frac{\alpha}{a} A, \qquad (F) = -\frac{\alpha}{a} F,$$

(11.1)

where α and a are the semimajor axes of the photocentric and the relative orbit respectively, and assuming constant blending effect β throughout the entire series of observations.

The following formulas are used for the orbital displacements in the two equatorial coordinates:

$$(B)x + (G)y \quad \text{in RA},$$

$$(A)x + (F)y \quad \text{in Decl.},$$

(11.2)

where x and y are the elliptical rectangular coordinates in unit orbit.

The fractional distance β of primary to photocenter in terms of the separation between the two components is given by

$$\beta = \frac{l_B}{l_A + l_B},$$

(11.3)

where l_A and l_B are the luminosities of the components or, in terms of magnitude difference Δm, companion minus primary, we have

$$\beta = (1 + 10^{0.4\Delta m})^{-1}.$$

(11.4)

The scale α of the *photocentric orbits* is therefore $(B - \beta)$ times that of the relative orbit, i.e. where B and β are the fractional mass and luminosity of the, assumed only, one companion:

$$\alpha = (B - \beta)a .\qquad\qquad (11.5)$$

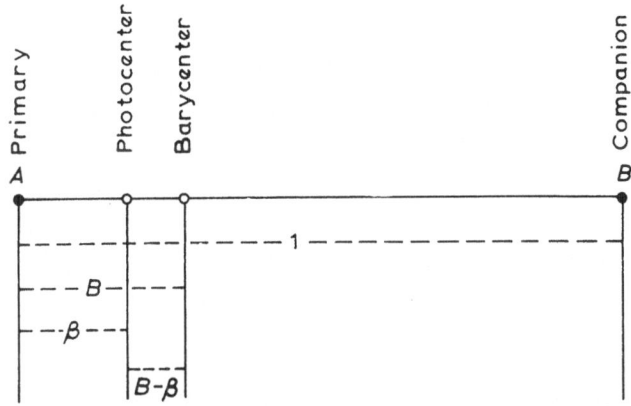

Relative spacing between components, barycenter, and photocenter for photographically unresolved binaries, assuming one companion.

It now becomes more explicit and elegant to use the following homogeneous equations of condition in which apparent and parallactic orbit appear, side by side, for the observed path of primary component or of photocenter

$$X = c_X + \mu_X t + q_X t^2 + \pi P_\alpha + \alpha Q_\alpha ,$$
$$Y = c_Y + \mu_Y t + q_Y t^2 + \pi P_\delta + \alpha Q_\delta .\qquad (11.6)$$

The last two terms in (11.6) represent the parallactic and the orbital displacements in the respective coordinates. The annual parallax factors P_α, P_δ represent the projected fractional coefficient in RA and Decl. of the unforshortened angular value (relative annual parallax) of one astronomical unit. The *orbital* factors Q_α, Q_δ are the corresponding projected fractional coefficients of the unforshortened angular value of the semi-axis major

$$\alpha = a_1 = Ba \quad \text{for primary (no blending)} ,\qquad (11.7)$$

$$\alpha = (B - \beta)a \quad \text{for photocenter} .\qquad (11.5)$$

The orbital factors are the projected values in right ascension (reduced to great circle) and in declination of the radius vector barycenter to primary or to photocenter, for unit orbit. The orbital factors are analogous to the parallax factors; the latter refer to the star's parallactic orbit, the former to the star's own apparent orbit.

The orbital factors are calculated from (the) known dynamical and orientational elements of the relative orbit:

$$Q_\alpha = (b)x + (g)y,$$
$$Q_\delta = (a)x + (f)y.$$

(11.8)

The *orientation* factors

$$(b) = -\cos\omega\sin\Omega - \sin\omega\cos\Omega\cos i,$$
$$(a) = -\cos\omega\cos\Omega + \sin\omega\sin\Omega\cos i,$$
$$(g) = +\sin\omega\sin\Omega - \cos\omega\cos\Omega\cos i,$$
$$(f) = +\sin\omega\cos\Omega + \cos\omega\sin\Omega\cos i,$$

(11.9)

are functions of the orientation elements ω, Ω, and i of the relative orbit of the two components, while x and y are the elliptical rectangular coordinates in unit orbit. The orientation factors are related to the Thiele–Innes constants as follows:

$$B = -(b)a,$$
$$A = -(a)a,$$
$$G = -(g)a,$$
$$F = -(f)a,$$

(11.10)

where a is the scale (semi-axis major) of the relative orbit of primary and companion.

D. Mass-Function. Orbital Constant. Dynamical Interpretation

The scale α is obtained in angular measure, seconds of arc, and is reduced to linear measure by dividing through the (absolute) parallax p, i.e.

$$\alpha = \frac{\alpha''}{p''}$$

(11.11)

now expressed in astronomical units of distance. As a rule the parallax p'' is known with adequate accuracy for the present purpose. For a 'resolved' visual double star with both components visible we have the harmonic relation

$$\mathcal{M}_A + \mathcal{M}_B = \frac{a^3}{P^2}.$$

(4.2)

If for an unresolved binary only the primary is separately recorded on the photographic plate, we know only that fraction α of a, which refers to the primary i.e.

$$\alpha = Ba,$$

(11.12)

where B is the fractional mass $\mathcal{M}_B(\mathcal{M}_A + \mathcal{M}_B)$ of the unseen companion in terms of the combined mass of primary and one companion. All we can evaluate is not the total mass but only the *mass-function*

$$B^3(\mathcal{M}_A + \mathcal{M}_B) = \frac{\alpha^3}{P^2} \tag{11.13}$$

which also may be written as

$$\mathcal{M}_B = \alpha P^{-2/3}(\mathcal{M}_A + \mathcal{M}_B)^{2/3} . \tag{11.14}$$

The basic dynamical information obtained from the observations and analyses is the quantity $\alpha P^{-2/3}$ named the *orbital constant*. Usually the determination of α is less precise than that of P. However, as in the case of visual binaries, a reduction in the error of the orbital constant may result due to a partial cancellation of the errors in α and in P. To obtain a value for $\mathcal{M}_B$ the most important aim of our orbital study, we must therefore adopt or estimate a value for $\mathcal{M}_A$, which may not be too difficult since absolute magnitude and spectrum of the primary are known.

The above formulae are valid for the case that the unseen companion does not contribute any light to the formation of the photographic image i.e. either because the companion is essentially dark, or in angular measure is sufficiently far away (say more than $2''$) from the primary. That a photographic image shows no companion does not exclude that nevertheless this image is the result of a blend of primary and a close companion which emits some light; however for a small angular separation the blended image or photocenter may still present a circular image. In this case we use the formula

$$\alpha = (B - \beta)a , \tag{11.5}$$

where β is the fractional luminosity $l_B/(l_A + l_B)$ of the companion in terms of the combined luminosity of primary and companion.

The mass-function formula now becomes

$$(B - \beta^3 (\mathcal{M}_A + \mathcal{M}_B) = \frac{\alpha^3}{P^2} \tag{11.15}$$

which also may be written as

$$\mathcal{M}_B = \alpha P^{-2/3} (\mathcal{M}_A + \mathcal{M}_B)^{2/3} + \beta(\mathcal{M}_A + \mathcal{M}_B) . \tag{11.16}$$

An adopted value of $\mathcal{M}_A$ now yields a lower limit for the mass $\mathcal{M}_B$ of the companion.

In general, to arrive at values for $\mathcal{M}_B$, and $\mathcal{M}_A$, one proceeds as follows. The period of revolution being known, a range in values of a is adopted, as well as a range value of β. This leads to several combinations of $\mathcal{M}_A$ and $\mathcal{M}_B$ values. Acceptable values are restricted since too large values of a or too small values of β might have led to earlier visual detection of the companion. These constraints generally put limits on acceptable values for the mass of the companion, and yield an upper limit for the luminosity.

Of special interest is the case of a very small value of the orbital constant $\alpha P^{-2/3}$. This would indicate an unseen companion of very small mass and low luminosity. To a high degree of approximation formula (11.16) may then be written as

$$\mathcal{M}_B = \alpha P^{-2/3}\, \mathcal{M}_A^{2/3} + \beta \mathcal{M}_A \tag{11.17}$$

or even

$$\mathcal{M}_B = \alpha P^{-2/3}\, \mathcal{M}_A^{2/3} \tag{11.18}$$

for the case of a substellar or a 'dark' planetary companion, in which case we have also

$$\mathcal{M}_A = \frac{a^3}{P^2}\, . \tag{11.19}$$

Note the increased chance of discovery of a perturbation, as measured by the scale α caused by a very small mass for a primary of small mass, such as a red dwarf, as compared with a primary of larger mass, such as our Sun. For a *very* small value of $\mathcal{M}_B$ (11.7) may be written as

$$\alpha = \frac{\mathcal{M}_B}{\mathcal{M}_A}\, a\, . \tag{11.20}$$

For several visual double stars, residuals from the orbit have revealed systematic deviations, which can be ascribed to a third, unseen, component close to one of the visible components. In such a case a combined analysis is made of the 'small' perturbation orbit and the large orbit of one visible component with respect to the center of mass of the other visible component and unseen companion. Such an analysis cannot determine to which of the visual components the unseen companion belongs. This can be determined only by measuring the orbital motion of the two visible components on a background of reference stars, astronomically or spectroscopically.

CHAPTER 12

PERTURBATIONS. EARLY MOSTLY PHOTOGRAPHIC STUDIES

A. General

In Chapter 11 we have mentioned the promising method of discovering unseen companions by astrometric means, either through an irregularity in the proper motion of a single star or in the orbital motion of a visual binary. The perturbations in these cases indicate the presence of an unseen companion (or companions), the method is particularly effective for long periods of revolution with corresponding larger amplitudes and for large differences in brightness between the primary and secondary component.

At present more than twenty well-established perturbations are known and attributed to unseen companions. The great majority of these prove to be red dwarfs on the lower end of the Main Sequence with masses above 0.085 solar mass, the critical lower limit for a *bona-fide* star. A striking exception is a still unseen companion, with a mass almost as large as the Sun; this, the fourth component in the multiple star Zeta Cancri most likely is a white dwarf or degenerate star.

Six unseen companions prove to have masses well below 0.085 solar mass. Five of these have masses ranging well above the mass of Jupiter, they are referred to as dark dwarfs. Two unseen companions appear to have masses less than that of Jupiter and must be regarded as planet-like objects (Chapter 15).

An early photographic discovery of a perturbation in a stellar path was that of Ross 614; the companion was later seen. The only other perturbation, which later led to the visual detection of the unseen companion was that of VW Cephei. This then is the current state of affairs besides the two classical discoveries of the by now visible companions of Sirius and Procyon (Chapter 9).

The first pre-photographic discoveries of the perturbations of Sirius and Procyon were made and announced by Bessel in 1844 (Chapter 9). We now also have the first two visual detections of the photographically discovered unseen companions of Ross 614 and VW Cephei. Further details about these two interesting objects are given in Chapter 14. That other well-established unseen companions have not yet been visually detected should not in the least discourage us, or deter us from further studies; the available evidence in all these cases clearly points to the difficulty of any visual detection. Obviously, and naturally, we should do everything in our power to aim for visual detection by whatever method, whether through 'surface' or by 'space' astrometry. There is reason to expect results from advancing techniques including infrared and space telescopes. Comments on this subject are found in 'Astrometric Search for Unseen Stellar and Sub-Stellar Companions to Nearby Stars and the Possibility of Their Detection', by Sarah Lee Lippincott (1978, *Space Sci. Rev.* **22**, 153–189). Future searches may thus eventually 'see' what are at present unseen companions, both bona-fide stars, but also dark dwarfs, even of planetary objects, as for example VB 8.

The first photographic perturbations were not the result of any planned programs but rather a by-product of parallax determinations. We refer to the discovery of the unseen companions of Ross 614 at *McCormick Observatory* by *Dirk Reuyl* (1906–1972) and of Mu Cassiopeiae and Alpha Ophiuchi at *Allegheny Observatory* by *Nicolas E. Wagman* (1905–1980). For the past half century systematic searches were carried out by the present author, first at the McCormick Observatory and since 1937 at the Sproul Observatory where stars generally within 10 parsecs and within reach of the 61-cm refractor are photographed on a regular basis, preferably each annual observing season and, for the very nearest stars, on several nights each year. The majority of stars on this program are red dwarfs; because of their relatively low masses the chances of finding a perturbation are much better than for a solar type star, although some of these have revealed perturbations as well. Several other observatories are now active in this field, notably Allegheny, McCormick and the United States Naval Observatory. Generally photographic discoveries and subsequent studies of perturbations require many years, if not decades of observations in order to separate the orbital motion caused by the unseen companion from the proper motion of the center of mass of the unresolved binary system. For a single star, a perturbation reveals itself as a deviation from uniform rectilinear motion, a 'variable' proper motion; for a double star, as an irregularity in Keplerian motion. To obtain a satisfactory perturbation orbit (i.e. the orbit of the primary or 'parent' star), around the center of mass or primary and unseen companion, all phases of the orbit should be satisfactorily covered. And this perturbation is i.e. elliptical and subject to the law of areas, with respect to the center of mass. The most important orbital elements, those which lead to the mass of the unseen companion, are the *scale* of the orbit i.e. the (unforshortened) semi-axis major α and the *period* of revolution P. The eccentricity of the orbit is of importance since it is relevant to the physical origin and character of the companion. The detailed orbital analysis has been discussed elsewhere (*Stellar Paths*, Chapter 13), it is in part repeated in Chapter 11 of this treatise.

B. Early Results

Over the past half century over 20 perturbations have been discovered photographically both in stellar paths and in double star orbits. The early discoveries at the McCormick and Allegheny observatories were followed by the intensive observing program of nearby stars begun in 1937 at the Sproul Observatory; since well over a decade the United States Naval Observatory is contributing in this field. The orbital periods discovered thus for range from close to one year up to some sixty years; longer periods most likely will be found, with continued observations. Of the eight perturbations discovered at the USNO six have short periods ranging from 0.87 to 7.2 yr. As long as the companion remains unseen, only a lower limit and estimated range of its mass can be obtained; an estimate of the luminosity can often be made.

The classical astrometric discoveries of unseen companions were presented in Chapter 9. The studies of Sirius and Procyon were based on fundamental star positions (see Chapter 1); the perturbations amounted to several seconds due to the combination of large parallax (0″378 for Sirius, 0″292 for Procyon) and large semi-axes major and periods of the perturbation: 2″3, and 50.09 yr for Sirius; 1″2, and 40.65 yr for Procyon). No further perturbations have been found in this manner.

The first photographic discovery of a perturbation, namely, of Ross 614, was made at the McCormick Observatory (1936, *Astron. J.* **45**, 133). The historical details of this particular discovery are presented in Chapter 14.

The second photographic perturbation was found at the McCormick Observatory in 1947, that of Mlb 377 by Harold Lee Alden (1890–1964), a pioneer in long-focus photographic astrometry. Mlb 377 is a nearby visual binary (parallax 0″100). Alden's attempt to determine the mass-ratio of this star resulted in the discovery of an appreciable perturbation. A subsequent study at the Sproul Observatory by John L. Hershey in 1973 led to the well-determined perturbation orbit of Mlb A, with $P = 15.95$ yr and $\alpha = 0″125$. The true companion has been tentatively detected by the IR speckle work in Arizona. With a provisional period of about 400 yr the mass-ratio of Mlb Aa and Mlb B is still rather uncertain.

Next to these first photographic perturbations (Ross 614 and Mlb 377) we record *three* subsequent discoveries under the leadership of Wagman at the Allegheny Observatory, of the photographic perturbation of Alpha Ophiuchi, Mu Cassiopeiae, and Ross 52. α Ophiuchi and Mu Cassiopeiae are intrinsically bright stars and therefore of particular significance, since so few perturbations have been found for stars of comparatively high luminosity. Details follow below.

ALPHA OPHIUCHI: 17^h34^m9, $+ 12°31'$ $m_v = 2.1$, *A5 III*

In 1946 Wagman announced a perturbation with a total deviation of 0″15 in a period of 9 years (1946, *Astron. J.* **51**, 209); two years later he confirmed this result (1948, *Astron. J.* **53**, 137). These studies were based on Allegheny plates obtained over the years 1918–1921 and 1934–1945.

A later study by Lippincott and Wagman based on combined Sproul and Allegheny material yielded $P = 8.5$ yr and $\alpha = 0″065$. For an adopted mass $M_A = 3 \mathcal{M}_\odot$, a minimum mass of $M_B = 0.6 \mathcal{M}_\odot$ is found. Most likely values of Δm indicate a Main-Sequence companion and greatest separation of 0″5 (in 1982).

See: S. L. Lippincott and N. E. Wagman: 1963, *Astron. J.* **68**, 283; 1966, *Astron. J.* **71**, 122.

MU CASSIOPEIAE: 1^h8^m3, $+ 54°52'$, $m_v = 5.12$, G5 VI

Wagman (1963, *Astron. J.* **68**, 352) discovered the perturbation with a period of 22 yr (Wagman *et al.*, 1963) for this population II subdwarf. Subsequently a study at Sproul yielded a period of 18.5 yr. An analysis made after periastron passage (1954.34) (Lippincott, 1980, *Publ. Astron. Soc. Pacific* **92**, 531) yielded $P = 21.43$ yr, $\alpha = 0″186 \pm 0″001$ (pe.) and a relative parallax of 0″130 $\pm$ 0″011. These were based on 215 nights over the interval 1937–1980. Several attempts have been made to detect the companion, made difficult became of its apparent faintness ($\Delta m \sim 4.5$). The best tentative values for the masses are $\mathcal{M}_A \simeq 0.68\,\mathcal{M}_\odot$, $\mathcal{M}_B \simeq 0.17\,\mathcal{M}_\odot$.

Ross 52: 14^h53^m9, $+23°34'$, $m_{pv} = 11.5$

Mrs Chrissman at Allegheny Observatory (1953 Dec., *Astron. J.* **58**, 239) found this perturbation. The star is about a cm off center on plates of BD $+ 23°2751$, which had been photographed with the Thaw refractor since 1927. 48 plates through 1952 measured in RA yield a parallax of $0\rlap{.}''112 \pm 0\rlap{.}''007$, allowing for perturbation. A period of 20.5 yr and half amplitude of $0\rlap{.}''09$ were found. The parallax of BD $+ 23°2751$ was found to be $0\rlap{.}''034 \pm 0\rlap{.}''006$.

Earlier Reuyl (*Publ. Astron. Soc.* **53**, 119, 1941) had found Ross 52 to be double on plates taken with the McCormick refractor: separation $0\rlap{.}''9 \pm 0\rlap{.}''1$, position angle $103° \pm 3°$, estimated visual $\Delta m = 0.6$; a relative parallax of $0\rlap{.}''082 \pm 0\rlap{.}''010$ was found.

As of 1973, Sproul measurements reveal an amplitude of 0.01 mm or about $0\rlap{.}''2$, mostly in RA and an estimated period of about 30.5 yr for this perturbation.

Zeta Cancri C: 8^h12^m6, $+17°39'$, 6.7 G2, $p = 0\rlap{.}''047$

The visual binary Zeta Cancri AB has a period of 59.7 yr and a semi-axis major of $0\rlap{.}''88$. The orbit of the distant companion Zeta Cancri C with respect to Zeta Cancri AB has a period of about 1150 yr and a semi-axis major $8\rlap{.}''0$; Zeta Cancri C, a solar type star (G2) has a well established perturbation with $P = 17.5 \pm 0.2$ yr, $\alpha = 0\rlap{.}''191$ (Gasteyer, 1954, O. Franz, 1958). This is the first classical discovery (1888) from micrometer observations of a perturbation in a visual binary (Seeliger, 1914). Assuming no light for the still unseen companion, its mass is found to be $0.9\,\mathcal{M}_\odot$. This fourth component of the by now quadruple system appears to be white dwarf.

Xi Ursae Majoris A: 11^h15^m5, $+31°49'$, 4.3 G0V, $p = 0\rlap{.}''130$

The visual binary Xi Ursae Majoris AB has a period of 59.9 yr and a semi-axis major of $2\rlap{.}''54$ (van den Bos, 1928), values not significantly changed by later orbit determinations. From micrometer observations of Xi UMa AB, Nörlund (1905) found this classical discovery of a perturbation in a binary, which had been noted by W. H. Wright (1900) from spectroscopic observations. The perturbation was confirmed from multiple exposure photographs by E. Hertzsprung (1919): $P = 1.8$ yr, $\alpha = 0\rlap{.}''052$. The unseen companion has an estimated mass of $0.3\,\mathcal{M}_\odot$. The fainter component Xi UMa B is a spectroscopic binary, thus making Xi UMa a quadruple system.

Zeta Aquarii B: 22^h28^m8, $+0°2'$, 4.6, F1, $p = 0\rlap{.}''043$

The long-period binary Zeta Aquarii AB was found to have a perturbation, analyzed from a combination of photographic and visual material (Strand, 1942). The unseen companion belongs to the B-component (Franz, 1958). A study by Harrington (1968) yields a period of 856 yr, a semi-axis major of $5\rlap{.}''055$ for the AB orbit, the corresponding values for the perturbation are 25.5 yr and $0\rlap{.}''072$; the mass of the unseen companion is $0.26\,\mathcal{M}_\odot$.

CHAPTER 13

PERTURBATIONS. CURRENT PHOTOGRAPHIC STUDIES

A. General

The information provided mostly by R. S. Harrington and by S. L. Lippincott is listed below in order of estimated mass of the companion upwards from 0.005 $\mathcal{M}_\odot$ but not more than 0.5 $\mathcal{M}_\odot$. For each star are given the position referring to the equator, equinox and epoch of the year 2000, the apparent visual magnitude, spectrum, parallax as well as the period of revolution P, eccentricity e of the perturbation orbit and the adopted value for the mass M_B of the unseen companion (Chapter 11, formulae (11.17ff)). Only for two stars, Ross 614 and VW Cephei the companion has been seen later (Chapter 14).

For several other stars duplicity has also been detected through infrared observations by Donald W. McCarthy Jr. and others. These and continued (visual) photographic studies will be followed with great interest.

B. Results

VB10: 19^h17^m0, $+5°6'$; 17.53 ± 0.01, M8 Ve, $B - V = 2.20 \pm 0''01$, parallax $0''178$. Commonly known as Van Biesbroeck's star (*Astron. J.* **66**, 528, 1961), the distant faint companion of Wolf 1055. VB10 perhaps is the star of best known lowest visual luminosity: $M_v = +18.6$ (Chapter 6).

87 plates taken over the interval 1972–1982 reveal a perturbation, for which

$$P = 4.9 \text{ yr}, \ e = 0.6, \ \alpha = 0''015 \ .$$

(R. S. Harrington, C. C. Dahn, and V. V. Kallavakal: *Astron. J.* **88**, 1038, 1983). From an extrapolation of the mass-luminosity relation (K. Aa. Strand, *Astron. J.* **82**, 1745, 1977), perhaps justified by the fact that VB10 appears to be on the extended Main Sequence in the color-luminosity diagram, the mass of the primary is estimated at 0.06 $\mathcal{M}_\odot$, and the mass of the secondary at 0.005 $\mathcal{M}_\odot$, i.e. five times that of Jupiter.

This is a most interesting and exceptional result for a very faint star. VB10 has to be considered a dark dwarf and is found to have an even much fainter dwarf companion, not yet resolved by infrared studies. Thus we seem to have an example of a double dwarf star. How many more are to be found?

A result like this is a tribute to the quality and power of the beautiful 155-cm reflector at the Flagstaff station of the U.S. Naval Observatory. The result could not have been obtained without the continued observational efforts which required long exposure times of one hour, and without patience, the *sine qua non* in the field of astrometric perturbations.

Two other perturbations leading to companions with mass below 0.01 $\mathcal{M}_\odot$ follow:

$$BD + 68°946, \ 17^h36^m4, \ +68°17' ; \ 9.2, \text{ M3V, parallax } 0''213 \ .$$

287 nights observations taken over the interval 1937–1976 yield:

$$P = 26.37 \text{ yr}, \ e = 0.90, \ \alpha = 0\rlap{.}''033 \pm 0\rlap{.}''002$$

(S. L. Lippincott, *Astron. J.* **82**, 925, 1977).

Note the rather high eccentricity. No elongation has been observed on the Sproul photographs, nor has the companion been observed visually or by infrared speckle studies by D. W. McCarthy.

Adopted mass of unseen companion 0.008 $\mathcal{M}_\odot$.

$BD + 43°4305$ (EV Lacertae): 22^h46^m8, $+44°17'$; 10.2, M4.5 e, parallax $0\rlap{.}''200$, 322 nights observations over the interval 1937–1979, indicate a perturbation with the following elements $P = 45$ yr, $e = 0.5$, $\alpha = 0\rlap{.}''0206 \pm 0\rlap{.}''0014$ (*Stellar Paths*, p. 113, 1981). Adopting a mass of 0.25 $\mathcal{M}_\odot$ for the primary, a mass of 0.009 $\mathcal{M}_\odot$ has been adopted for the companion. Don McCarthy has not resolved this star.

Stein 2051 A: 4^h31^m2, $+58°56'$; 11.1, M5, $p = 0\rlap{.}''181$, $P = 23$ yr, $e = 0.3$, $\alpha = 0\rlap{.}''07$

$$M_B = 0.018 \ \mathcal{M}_\odot .$$

CC 1228: 20^h45^m1, $+44°27'$; M3, $p = 0\rlap{.}''082$, $P = 6.3$ yr, $e = 0.5$, $\alpha = 0\rlap{.}''012$

$$M_B = 0.022 \ \mathcal{M}_\odot .$$

VB8: 16^h55^m6, $-8°20'$; 16.8, M, $p = 0\rlap{.}''154$, $P > 60$ yr, $e = \ldots ?$

$$M_B = 0.03 \ \mathcal{M}_\odot .$$

G139–29: 17^h17^m7, $+11°37'$; 15.1, M, $p = 0\rlap{.}''080$, $P = 10$yr, $e = 0.2$

$$M_B = 0.04 \ \mathcal{M}_\odot .$$

G96–45: $.5^h33^m3$, $+44°46'$; 12.2, MD, $p = 0\rlap{.}''066$, $P = 7.2$ yr, $e = 0.01$

$$M_B = 0.05 \ \mathcal{M}_\odot .$$

G107–69: 7^h30^m7, $+48°9'$; 13.3, M, $p = 0\rlap{.}''091$, $P = 0.85$ yr, $e = 0.7$

$$M_B = 0.06 \ \mathcal{M}_\odot .$$

G24–16: 20^h29^m8, $+9°38'$; 13.0, M, $p = 0\rlap{.}''114$, $P = 1.5$ yr, $e = 0.45$

$$M_B = 0.06 \ \mathcal{M}_\odot .$$

C986: 16^h24^m2, $+48°18'$; 10.3, M0, $p = 0\rlap{.}''134$, $P = 3.72$ yr, $e = 0.53$

$$M_B = 0.06 \ \mathcal{M}_\odot .$$

Ross 614: 6^h29^m4, $-2°46'$; 11.7, M4, $p = 0\rlap{.}''248$, $P = 16.6$ yr, $e = 0.38$

$$M_B = 0.06 \ \mathcal{M}_\odot .$$

VW Cephei: 20^h37^m4, $+75°33'$; 7.2, G8, $p = 0\rlap{.}''041$, $P = 30.4$ yr, $e = 0.60$

   $M_B = 0.06 \, \mathcal{M}_\odot$.

36 UMa: 10^h30^m6, $+55°56'$; 4.8, F8V, $p = 0\rlap{.}''079$, $P = 18$ yr, $e = 0.8$

   $M_B = 0.07 \, \mathcal{M}_\odot$.

G146–72: 10^h55^m1, $+47°12'$; 12.7, M, $p = 0\rlap{.}''033$, $P = 6.7$ yr, $e = 0.2$

   $M_B = 0.09 \, \mathcal{M}_\odot$.

BD + 6°398: 2^h36^m1, $+6°50'$; 5.8, K3V, $p = 0\rlap{.}''128$, $P = 50$ yr, $e = 0.60$

   $M_B = 0.12 \, \mathcal{M}_\odot$.

Wolf 922: 21^h31^m3, $-9°45'$; 12.0, M4, $p = 0\rlap{.}''122$, $P = 1.93$ yr, $e = 0.36$

   $M_B = 0.12 \, \mathcal{M}_\odot$.

χ′ Orionis: 5^h54^m4, $+20°13'$; 4.4, G5VI, $p = 0\rlap{.}''102$, $P = 14.25$ yr, $e = 0.6$

   $\mathcal{M}_B = 0.17 \, \mathcal{M}_\odot$.

Wolf 1062: 19^h12^m2, $+2°51'$; 11.1, M4, $p = 0\rlap{.}''101$, $P = 2.3$ yr, $e = 0.5$

   $M_B = 0.18 \, \mathcal{M}_\odot$.

BD + 27°4120: 21^h38^m0, $+27°40'$; 9.8, M0, $p = 0\rlap{.}''077$, $P = 85$ yr, $e = 0.3$

   $M_B = 0.20 \, \mathcal{M}_\odot$.

BD + 67°552: 8^h36^m4, $+67°15'$; 9.3, M1, $p = 0\rlap{.}''070$, $P = 23$ yr, $e = 0.73$

   $M_B = 0.30 \, \mathcal{M}_\odot$.

BD + 41°328′: 1^h41^m8, $+42°34'$; 5.0, G1V, $p = 0\rlap{.}''070$, $P = 19.5$ yr, $e = 0.42$

   $M_B = 0.38 \, \mathcal{M}_\odot$.

G107–70: 7^h30^m8, $+48°8'$; 14.4, WD + WD, $p = 0\rlap{.}''091$, $P = 20.5$, $e = 0.18$

   $M_A = M_B = 0.46 \, \mathcal{M}_\odot$.

C. Review

The majority of perturbations are found on the lower end of the Main Sequence (Chapter 6). Five perturbations point to unseen companions with masses M_B from 0.005 to 0.022 $\mathcal{M}_\odot$. The eccentricities of the perturbation orbits range from 0.3 to 0.9, averaging 0.56. Seven more suggest companions with $\mathcal{M}_B$ ranging from 0.03 to 0.06 $\mathcal{M}_\odot$ and orbital eccentricities averaging 0.47.

Compared with the adopted value $\mathcal{M}_A$ for the visible primary, $\mathcal{M}_B$ is always less than 10% of $\mathcal{M}_A$. In our solar system the mass-ratio planet-Sun is always less than 0.1%, the largest value is 1/1047 for Jupiter; the mass-ratios seem extreme, compared with those of conventional double stars. The orbital eccentricities of the latter average 0.6 (small eccentricities are lacking), very comparable with the perturbation orbits.

For the planets of the Sun the largest orbital eccentricities are 0.206 for Mercury and 0.248 for Pluto; both have large orbital inclinations, 7° and 17°, with respect to the ecliptic. For the Jupiter-planets, the orbital eccentricities average 0.02, being largest, 0.056 for Saturn. Perhaps the planetlike companions of Barnard's star have small orbital eccentricities (Chapter 15).

We repeat (Chapter 8): planets appear to originate from a *disk*, resulting in co-planar and co-revolving orbits of small eccentricity. *Double star components* do not come from a disk.

TWO CASES OF SERENDIPITY: ROSS 614 AND VW CEPHEI

> Serendipity: from the Concise Oxford Dictionary, fourth edition 1951 "faculty of making happy and unexpected discoveries by accident" coined by Horace Walpole after *The Three Princes of Serendip* (Ceylon), a fairy tale.

Progress in science depends to a great extent on planed programs of research. However we must not ignore the role played by serendipity. Significant discoveries now and than are made unintentionally, as an unexpected byproduct, of a program planned for a different purpose. A classical example is the discovery of stellar aberration in 1726 by James Bradley. Intent on finding stellar parallax, which later proved to be below Bradley's attainable accuracy, he did find the yearly aberration effect, differeing 90° in phase from parallactic displacement, with a total amplitude of 41″ (Chapter 3).

A. Ross 614, a Case of Quadruple Serendipity

We shall now report the story of quadruple serendipity, due to the happy cooperation of four different happenings: initial inaccuracies, personal interference and twice: lucky timing. It is the early history of the now well-known visual binary Ross 614. The serendipity sequence is ended by a climax, namely the sighting and photographing of the fainter component of Ross 614, the intrinsically faintest visible star with smallest well determined mass.

In the nineteen twenties the determination of stellar parallaxes by the technique and methods of long-focus photographic astrometry was one of the foremost activities of observational astronomy. Since only a limited number of stars could be observed, the selection of stars for parallax programs was, and still remains, a basic problem. While systematic studies of the naked eye stars have also been pursued, large proper motion stars were and still remain, a principal source for the planning of parallax programs. A good choice, since generally these objects prove to be comparatively near. Parallax programs have thus relied heavily on the searches for appreciable proper motions by Max Wolf (1863–1932), F. E. Ross and more recently those of W. J. Luyten, H. L. Giclas, and others.

PROPER MOTION DISCOVERIES BY ROSS

We borrow from Seth B. Nicholson's excellent biography (*Publ. Astron. Soc. Pacific* **73**, 182, 1961). Frank Elmore Ross's (1874–1960) initial interest lay in the field of celestial mechanics; he worked for a year with Simon Newcomb at the Nautical Alamanic Office in Washington, D.C. From 1905 till 1915 he was director of the International Latitude Observatory in Gaithersburg, Pennsylvania. From 1915 to 1924 he was a physicist in the laboratories of the Eastman Kodak Company in Rochester, New York State. There he dealt with the techniques of photography and with lens design. He wrote a classical treatise on the Physics of the photographic image (1924).

In 1924 Ross joined the staff of the Yerkes Observatory in Williamsbay, Wisconsin. He re-photographed stellar fields, that had been covered earlier, over the years by E. E. Barnard with the Bruce photographic doublet of 25 cm aperture and 128 cm focal length. Ross thus discovered many variable stars and proper motion stars. Over the years 1926–1931 he published proper motions of 869 stars, generally between photographic magnitudes 8 and 14, and with proper motions exceeding $0\overset{''}{.}2$ annually. The proper motions were detected with the Zeiss stereo comparator using the blink attachment.

Ross retired in 1939 to Pasadena, California, where the maintained close connection with the Mount Wilson Observatory. He continued his work on designing and computing lenses. 'Ross-lenses' in various places, including the 50 cm astrograph at the Lick Observatory, are a lasting tribute to him.

At the Leander McCormick Observatory, Alexander Nicolavitch Vyssotsky (1888–1973) and the author followed these Ross discoveries with considerable interest and discussed with director Samuel Alfred Mitchell (1874–1960) the desirability of adding a number of these Ross stars with comparatively large proper motion to the parallax program of the 66 cm McCormick refractor. The fainter Ross stars generally required comparatively long exposure times, however Dr Mitchell magnanimously though reluctantly agreed to a compromise which included stars not fainter than 11 pg and with proper motions exceeding 1″ annually. Star No. 614 was listed by Ross (1926, *Astron. J.* **862**, Vol. 37, p. 195) with a magnitude 11 and total proper motion $1\overset{''}{.}09$ in position angle 132°. Thus it happened that the star just made it and was put on the McCormick parallax program in 1927.

FIRST McCORMICK PARALLAX DETERMINATION OF ROSS 614

The first plate taken on October 24, 1927 with two images had exposure times of 25 min; the following two plates taken on March 2, 1928 had exposure times of 15 min. After that the exposure time generally remained above 10 min; it was not until 1938 that the exposure time dropped below 10 min; beginning 1943 the exposure times went below 5 min; since 1968 the basic exposure time was 2 min of time.

Meanwhile Carl L. Stearns had determined a relative parallax of $+0\overset{''}{.}250 \pm 0\overset{''}{.}012$ with the 51 cm refractor of the Van Vleck Observatory in Middletown, Connecticut (*Astron. J.* **44**, 152, 1935; *Publ. Van Vleck Observatory* **1**, 114, 1938). Many years later, the McCormick parallax series was completed and Ross 614 was found to have a relative parallax $+0\overset{''}{.}262 \pm 0\overset{''}{.}008$ (p.e.). The current adopted value of the absolute parallax is $+0\overset{''}{.}234$. The star thus proved to have been a good candidate for proximity to our solar system.

But, surprise, the apparent visual magnitude was found to be 11.3 and later evidence showed the proper motion to be less than the value 1″.09 given by Ross. Sproul photographic studies gave 0″.99 (Lippincott, 1950; Lippincott and Hershey, 1972) while the McCormick results (Probst, 1977) hover dangerously close to 1″.00. A Cincinnati analysis (Porter, Yowell, and Smith) gave 0″.97 for the total proper motion (*Astron. J.* **44**, 132, 1935). How fortunate that the magnitude and proper motion measured by Ross were sufficiently inaccurate, in the right direction, so that the star was considered eligible for the McCormick proper motion program! Had Ross estimated the photographic magnitude at 12, or 13, as established later, and had he found say 0″.95 or even ″.99 for the proper motion, the star very likely would have had to wait several years, before having been put on the McCormick parallax program. But there was no undue delay; from the directional restraints on proper motion and brightness the star had got in just 'under the wire' and immediately had been put on the McCormick parallax program.

These two pieces of luch were followed by two more. A typical parallax series in those days consisted of a series of some twenty plates, taken during 5 successive seasons covering an interval of two years. There were of course deviations from this ideal scheme, directed by the weather and fate ('condition humaine'). Ross 614, because of its comparative faintness and longer exposure time ($\sim$ 10 min) was not given highest directional priority and had to struggle in the competition for getting plates taken. And it therefore took a timespan of as long as 8 years, 1927, October 24–1935, October 8, before the parallax series consisting of 25 plates taken on 13 nights was considered to be completed. In retrospect: how wonderful!

DISCOVERY OF THE PERTURBATION OF ROSS 614. FIRST ORBITS

The measuring of the 'parallax' plates on Ross 614 was assigned to staffmember Dirk Reuyl. As was customary in those days the plates were measured in right ascension only, since most of the annual parallactic displacement is in that coordinate. This is particularly true for Ross 614. At its location (RA $6^h26.8$, Decl. $-2°46'$, 1950) the parallax factors for this early McCormick series ranged from $+0.98$ to -0.98 in RA, and only from -0.04 to -0.38 in Decl.

I clearly remember the day when Reuyl had made a least squares solution for proper motion and parallax of Ross 614 and noticed and showed us the striking nonlinear run in the residuals, which appeared to point to a perturbation. I suggested that Reuyl immediately measure the plates in declination. After having done so and after reducing the measures, the results revealed an even larger perturbation in declination! Great happiness, although it took some time before it dawned upon all of us, that this was the *first perturbation*, discovered by the precision approach of long-focus photograph astrometry, the same technique which had proven so successful in parallax determinations.

In the first parallax determination the deviations from uniform proper motions could be represented by a quadratic time effect of $0″.0019 \pm 0″.0013/\text{yr}^2$ and $-0″.0210 \pm 0″.0015/\text{yr}^2$ in RA and Decl. respectively. Adding 9 more plates taken on 5 nights the total material was increased to 34 plates, on a total of 18 nights, over the interval 1927, October 24–1936, March 15. Reuyl now derived a provisional period of

9 years for the perturbation and suggested a separation of something like $0''.8$ for the invisible companion. In the absence of any visual detection, the companion was assumed to be considerably fainter than the primary; this later on proved to be the case. The observing data and the results were published by Reuyl in the now classical paper, carrying the modest title 'Variable Proper Motion of Ross 614', 1936, *Astron. J.* **45**, 133.

Obviously considerable credit for this exciting discovery was due to Dr Mitchell, who had not given Ross 614 a chance to have its parallax determined in two years. By frowning upon this faint newcomer with long exposure times, and giving it low observing priority, Mitchell unknowingly had made it possible for the perturbation to reveal itself. Over an interval of two years, there would have been no measurable perturbation, but the extended interval of eight years did it. Roughly, over a fraction of its period, *ceteris paribus*, the weight of the perturbation effect ('curvature' or 'acceleration') increases with the fifth power of the time (*Stellar Paths*, 1981, 45ff). Still an additional piece of luck, which became evident later. The perturbation effects over the interval 1927–1936, were particularly large because periastron of the perturbation orbit occured in 1932.9, around the middle of the plate series! Hence the orbital effect, over the interval of the series was optimal, in both coordinates.

Although in the normal course of events, the perturbation of Ross 614 eventually would have been discovered, how fortunate that serendipity entered the picture, and thus gave us early information about this most interesting object, the star of low mass apparently below the critical value $0.085 \, \mathcal{M}_\odot$.

A second McCormick parallax determination based on 26 plates taken on 14 nights over the short interval 1935, October 27–1937, December 3, and requiring no quadratic time terms, yielded a relative parallax of $+ 0''.256 \pm 0''.011$ (*Publ. McCormick Observatory* **8**, 210–211; *Astron. J.* **50**, 117).

In 1941 Reuyl presented a new analysis of the McCormick material now extending over 13 years. Approximating the motion of the barycenter from a comparison with Algiers and Harvard positions of 1896 and 1905 respectively, the most probable value of the period was now found to be 14 years, the photocentric semi-axis major $0''.28$. Also determinations of the magnitude of Ross 614 were found to be 11.4 pv and 12.5 pg respectively and a spectral classification of M2e by A. N. Vyssotsky. The mass of the companion was found to be less than $0.1 \, \mathcal{M}_\odot$. Meanwhile Carl Stearns had confirmed the perturbation from photographic observations with the 51 cm refractor of the Van Vleck Observatory (*Publ. Am. Soc.* **10**, 27, 1939).

SUBSEQUENT ORBITAL STUDIES. VISUAL AND PHOTOGRAPHIC DETECTION OF ROSS 614 B

The case of serendipity has now been presented. The subsequent story of Ross 614 is well known; it is now a regular resolved binary consisting of two stars at the lower end of the Main Sequence, and therefore of particular interest. We record a brief review of some of the subsequent developments and studies. In 1950 Sarah Lee Lippincott published a next well determined orbit for the perturbation of Ross 614. The material included 309 plates taken on 92 nights taken with the Sproul 61-cm refractor over the interval 1938–1950, combined with earlier published McCormick material (*Astron. J.* **55**, 236, 1950). Lippincott's analysis yielded a period of 16.5 yr and a photocentric semi-axis major of 0".306 ± 0".006, both values not differing too much from Reuyl's results. The relative parallax was found to be + 0".247 ± 0".002, the total proper motion 0".991 in P.A. 134°. She could predict that greatest separation between the visible and the as yet still unseen component of the unresolved binary Ross 614 would take place around 1955. She requested Walter Baade (1893–1960) to examine the star visually with the 5-m Hale reflector, Baade had four nights during the month of February with this telescope at his disposal. The weather was terrible. Three more chances in March, poor seeing on the first two nights, on the third night March 23 conditions were favorable, Baade saw and immediately photographed, the companion with exposure times ranging from one to ten seconds. The companion was estimated to be 3.5 magnitudes fainter than the primary. The position angle was found to be 36° ± 2°, in excellent agreement with the value 37°, 5, predicted from the perturbation orbit. The separation measured on the photographic plate proved to be 1".19 with uncertanty due to photographic. For further details see Lippincott's articles (*Astron. J.* **60**, 379, 1955; and *Sky Telesc.* **364**, 1955).

A later analysis by Lippincott and Hershey (*Astron. J.* **77**, 679, 1972) was based on 868 plates taken with the Sproul refractor on 252 nights covering the interval 1938.9–1972.0. The companion has now been seen and its location as well as its magnitude measured by several observers (Lippincott and Hershey, 1972).

R. G. Probst made an analysis (*Astron. J.* **82**, 656, 1977), based on 285 plates taken over the interval 1928.2–1975.9 with the McCormick refractor on 142 nights.

The principal results of all these exhaustive investigations are the determinations of the mass of the companion, which turns out to be 0.06 $\mathcal{M}_\odot$ with an estimated uncertainty of 0.01 $\mathcal{M}_\odot$, possibly more.

We keenly realize that the present example is but an exceptional event and the bulk of hidden treasures in observational research is found by careful long-term planning. Meanwhile we remain grateful to fate for the occasional significant contribution engendered by serendipity.

B. The Discovery of VW Cephei C – Serendipity Again

INTRODUCTION

It seems appropriate to add at this stage another case of serendipity. VW Cephei is a W Ursae Majoris type eclipsing binary ($20^h37.4 + 75° 33'$, 2000) with a period of 0.278 days. The variability was discovered photographically in 1926 by Jan Schilt (1894–1982) at the Mount Wilson Observatory. Times of minimum have been observed since 1890. Early photometric studies include those of H. van Gent and R. S. Dugan. The apparent visual magnitude ranges between 7.3 and 7.7, the system consists of two solar-type stars (spectra G5 and K) gravitationally elongated because of their being so close together.

The period is not constant, both maxima and minima show a variation with a semi-amplitude of (0.065 times the period) about half an hour, over a period of 26 years. The star was studied by Cecilia Payne-Gaposchkin (1901–1979). Biographies of this brillant astronomer have been written by Charles A. Whitney (*Sky Telesc.* **59**, 212, 1980) and by Owen Gingerich (*Quart. J. Roy. Astron. Soc.* **23**, 450, 1982). She proposed that the deviation from a constant period is a light-time effect due to an orbit around a third invisible component. The third body would be almost indentical with the other two in mass, luminosity and spectrum (*Publ. Am. Astron. Soc.* **10**, 127, 1941). The deviations of the times of minima, continued to be interpreted as a light-time effect by H. Schmidt and K. W. Schrick (1955), by T. Herczog and H. Schmidt (1960), who propose an underluminous third component.

SPROUL ASTROMETRIC AND PHOTOMETRIC STUDY

After hearing Cecilia Payne-Gaposchkin's fascinating presentation at the 65th meeting of the American Astronomical Society in Philadelphia in December 1941, I returned to Swarthmore, and immediately put VW Cephei on the observing program of the Sproul telescope. Now, more than four decades later I am very glad I did. Obviously a star, in this case the very close binary VW Cephei, with a suspected unseen companion, belonged on the Sproul program, which was dedicated to finding unseen companions, generally of stars within ten parsecs. True, VW Cephei was not that near to us, but appeared to have a third unseen component of sufficient mass to cause a measurable perturbation, and deserving therefore was a promising candidate for the Sproul astrometric program.

What follows now, is quoted from the exhaustive study made by John L. Hershey (*Astro. J.* **80**, 662, 1975; also *Sky Telesc.* **50**, 230, 1975). A preliminary study of the Sproul astrometric series was made by J. F. Villamediana (1978, Masters thesis Swarthmore College, unpublished).

At first it looked as if the astrometric data seemed consistent with a light-time hypothesis, but then the third component should have been visually detectable. Using all astrometric data over the interval 1942–1974, 610 plates taken on 164 nights with the Sproul 61-cm refractor, Hershey made an astrometric analysis, independent of photometric data. The relevant results were as follows:

> Relative parallax + 0″0394 ± 0″0020
> Absolute parallax + 0″041 ± 0″002
> Period 30.45 ± 0.17 yr
> eccentricity 0.494 ± 0.028
> Periastron 1966.48 ± 0.20
> semi-axis major of perturbation 0″130 ± 0″002 = 3.2 ± 0.2 AU.

Next Hershey made a photometric study, using available times of minima, 450 in all, from 1926 to 1974.

Using a period of 0.278 318 3 days, it appeared that what had been believed to be a light time effect, was inconsistent over the entire interval, with the astrometric predictions. The conclusion had to be reached that the *true period changes* had taken place in the close eclipsing binary. Moreover the astrometric data, and with the help of a time-or-minimum ephemeris, W. D. Heintz visually examined VW Cephei in the summer of 1974, saw the companior with the Sproul refractor and made the first visual measurement with the filar micrometer of the by now 'seen' companion of VW Cephei. The observation differed only one degree from the position angle predicted from the astrometric orbit and agrees well with the separation from the spectroscopic mass of VW Cephei AB and the astrometric orbit. The mass of VW Cephei C was found to be $0.58 \pm 0.14 \, \mathcal{M}_{\odot}$, the absolute visual magnitude 8.2, the masses of VW Cephei A and B are $1.1 \, \mathcal{M}_{\odot}$ and $0.4 \, \mathcal{M}_{\odot}$, respectively.

Again a serendipitous case. Paradoxically the behaviour of light time residuals which prompted the long-term astrometric survey, has turned out to be unrelated to a third component, being caused by a combination of both sudden and gradual changes in the period of VW Cephei AB. This star at a distance over 20 parsec had been put on the observing program for the wrong reason. But VW Cephei has obliged; the astrometric efforts nevertheless resulted in the discovery of a bona-fide companion for a star at a distance over 20 parsec, well beyond the limit for any 'planned' discoveries of unseen companions to nearby stars. How many similar situations do exist?

CONCLUSION

Both the red dwarf companions of Ross 614 and of VW Cephei, owe their early discovery to the cooperation of serendipity. It should be mentioned also that of the dozens of unseen companions discovered photographically over the past several decades only those of Ross 614 and VW Cephei have been visually detected so far. A mere coincidence, I suppose, unless we invoke the additional serendipity of fate entering the picture.

While Ross 614 B represents the very lowest end of the Main Sequence, VW Cephei is located higher up, not far from the Sun. This would seem to hold out hope for visual discovery of the many other unseen companions found thus far but not yet seen.

CHAPTER 15

ASTROMETRIC STUDY OF BARNARD'S STAR

A. Proper Motion, Parallax and Acceleration

At present it is not possible to 'see' a planet as large as Jupiter and shining by reflected light as Jupiter does, even from the nearest solar type star, the brighter component of the Alpha Centauri system at a distance of 4.3 lightyears. What nes techniques may bring, remains to be seen. Meanwhile the perturbation appproach exists and is promising particularly if a 'heavy' Jupiter-like planet, or planets, were orbiting a nearby star of low mass i.e. a red dwarf. A planet with the mass of Jupiter orbiting Alpha Centauri A (parallax 0″.76) with a period nearly 12 years and semi-axis major of 5 astronomical units, a situation analogous to the Sun-Jupiter situation, would cause a perturbation with a total range of 0″.0076. The same planet orbiting nearby Barnard's star (parallax 0″.55) in the same size orbit, and correspondingly larger period of revolution of nearly 32 years would cause a much larger perturbation with a total range of 0″.04, since Barnard's star has only 1/7 times the mass of Alpha Centauri A. Were Barnard's star as near as Alpha Centauri, the total perturbation range would be 0″.054.

As we have seen in chapter 3, the attainable acuracy in long-focus photographic astrometry is something like 0″.02 for the relative position of two star images. An error as low as 0″.005 may be reached for the average of several exposures on a number of plates taken on any one night for the position of a star relative to a background of reference stars. An even higher acuracy may be built up for the average of data obtained over several nights during an observing season; errors as small as 0″.002 have been attained.

There exists therefore the real possibility of detecting perturbations for nearby stars caused by Jupiter-like companions, with periods of the order of more decades. The discovery of perturbations caused by planetary companions with Earth-like masses, must be considered as impossibility at this time.

The intensive and extensive studies of the path of Barnard's star, carried out at the Sproul Observatory over more than four decades appear to have resulted in the discovery of two Jupiter-like companions. A report on this interesting star will now be given; see also Vistas in Astronomy **26**, 144, 1982.

Barnard's star, an ordinary red dwarf, of particular interest because it is next to the Sun and Alpha Centauri system the nearest known star, was discovered in June 1916 by the well-known amateur astronomer Edward Emerson Barnard (1857–1923) from its larger proper motion of 10″.31 annually. At this rate its displacement on the sky equals the size of the Moon's diameter in about 180 years. The large proper motion, a sign of nearness, led to parallax-determinations by several observatories; the parallax is now known with considerable accuracy: 0″.547 ± 0″.003 (p.e.) corresponding to a distance of 1.83 parsecs or 6.0 lightyears. Observations of the Doppler shift in the

312

spectrum give a radial velocity of approach 108 km s^{-1}. The velocity perpendicular to the line of sight – the tangential velocity – is found to be 89 km s^{-1}. The resultant space velocity (relative to the Sun) is 140 km s^{-1}.

Barnard's star, a member of the population of our stellar neighborhood partakes in the general rotation of our Milky Way System (Chapter 2). The huge time scale of this galactic rotation prevents the detection of galactic orbital effects i.e. 'curvature' in the paths of individual stars for millenia to come. Moreover, the stars in our neighborhood are spaced so far apart that no measurable mutual gravitational effects result. Thus for a long time to come, hundreds if not thousands of years, individual stellar space motions as well as the space motions of barycenters of binary stars, when referred to the Sun, or rather the berycenter of the solar system may be considered non-accelerated i.e. rectilinear at constant velocity apart from any perturbations due to unseen companions.

In an astrometric study of Barnard's star the perspective acceleration, must be recognized and eliminated, from any measurements, lest it contaminates any perturbation.

We give the general formulae for the secular changes:

Annual change in proper motion μ

$$\frac{d\mu}{dt} = -2\overset{''}{.}05 \times 10^{-6} \, \mu p V$$

Annual change in parallax p

$$\frac{dp}{dt} = -1\overset{''}{.}02 \times 10^{-6} \, V p^2$$

Annual change in radial velocity V

$$\frac{dV}{dt} = +2.30 \times 10^{-5} \frac{\mu^2}{p} \, \text{km s}$$

Annual change in apparent magnitude m

$$\frac{dm}{dt} = +2.17 \times 10^{-6} \, V p$$

For Barnard's star

$$V = -108 \text{ km s}$$
$$\mu = 10\overset{''}{.}31 \text{ yearly}$$
$$p = 0\overset{''}{.}547$$

Hence the following current values for the secular changes

$$\frac{\mathrm{d}\mu}{\mathrm{d}t} = + 0\rlap{.}''00125 \ \mathrm{yr}^{-2}$$

$$\frac{\mathrm{d}p}{\mathrm{d}t} = + 0\rlap{.}''000032 \ \mathrm{yr}^{-1}$$

$$\frac{\mathrm{d}V}{\mathrm{d}t} = + 0.0046 \ \mathrm{km \ s \ yr}^{-1}$$

$$\frac{\mathrm{d}m}{\mathrm{d}t} = - 0\rlap{.}''\!0013 \ \mathrm{yr}^{-1}$$

Thus the proper motion of Barnard's star, $10\rlap{.}''31$ annually, presently is increasing at the annual rate of $0\rlap{.}''0015$; this rate will increase to $0\rlap{.}''00242$ about AD 5.900, then decrease to zero about AD 11.800 at which epoch the proper motion will reach a maximum value $25''$.

The parallax now $0\rlap{.}''547$, distance 6.0 lightyears, presently increases at the rate of $0\rlap{.}''000032$ annually i.e., the distance decreases at the annual rate of 0.000 36 lightyears. This rate increases to $0\rlap{.}''000034$ and 0.000 38 lightyears respectively about AD 6.100, decreases to zero about AD 11.800, when the parallax reaches its maximum value of $0\rlap{.}''85$; the closest approach will be 3.85 lightyears. At closest approach the star will appear in the constellation Draco, $51°$ degrees to the north from its present direction.

The radial velocity, now $- 108 \ \mathrm{km \ s}^{-1}$, presently increases at the annual rate of $0.0046 \ \mathrm{km \ s}^{-1}$; this rate increases to a maximum of $0.0120 \ \mathrm{km \ s}^{-1}$ at closest approach.

The apparent visual magnitude 9.5, presently decreases at the annual rate of 0.000 13 magnitude; this rate increases to 0.00015 magnitude about AD 3.600 and decreases to zero about closest approach AD 11.800 when the apparent brightness has increased to maximum apparent magnitude 8.5.

The time has come to take into account secular changes in proper motions and eventually also in parallaxes and radial velocities. These changes are not of a dynamical nature; they are secular perspective effects resulting from changing distance and angle of view. In other words, if the space velocity remains constant, proper motion, radial velocity and parallax change as time goes by. At the same time these changes themselves are subject to secular changes. The perspective change in proper motion, by far the most important secular effect, is proportional to the proper motion, the parallax, and the radial velocity, all three of which have appreciable values for Barnard's star.

The values of the secular changes with time, are symmetrical with respect to closest approach, except for a change in sign for $\mathrm{d}\mu/\mathrm{d}t$, $\mathrm{d}p/\mathrm{d}t$, and $\mathrm{d}m/\mathrm{d}t$. For further details see *Stellar Paths*, Chapter 8. Of these secular changes only the secular perspective acceleration plays a role at present: its accumulated effect with time is sufficiently large to be measurable and to be allowed for in any perturbation study.

B. Astrometric Profile of Sproul Refractor

The reality of the perturbation in the observed path of Barnard's Star, or of any other star, may be tested by establishing the astrometric profile of the telescope over the years. Ideally this is obtained from the paths of stars without any measurable perturbation i.e. 'single' stars, or from the barycentric paths of either resolved or unresolved astrometric binaries. Allowing for parallax and proper motion, and for orbital motion and quadratic time effect if need be, the residuals thus yield the astrometric profile of a 'composite' star path. Fortunately a fairly strong material is available; twenty-three stellar paths have been used with a total weight of 7880 (3110 nights), compared with a total weight of 3800 (1200 nights) for Barnard's star. New weights (Lippincott, 1974) are used from here on.

The majority of stars on the Sproul observing program are nearby red dwarfs, and it is among these, what we first look for candidats to determine the astrometric profile of the Sproul refractor. Several of these objects have had continued observational coverage for four decades or more since 1937; they include 10 suitable series of stars with spectral types K7 to M5, namely:

BD + 5° 1668, Wolf 294, Groomb. 1618, Ross 128,
Σ 2398, Σ 3121, ADS 10585, CC 986, CC 1299, and Furuhjelm 54.

These stars were studied by Lippincott, Hershey and associates (1975, 1978, 1980, 1982). Observations over the interval 1937–1980 were made on a total of 1740 nights, and represent a total night weight of 4950. Of the 10 stars, the first four, representing 56% of the total weight of the group, are 'single', thus far showing no detectable perturbation. The most powerful contributor (24%) is BD + 5° 1668, which, by itself, confirms the essential stability of the Sproul telescope over four decades (Lippincott and Hershey 1982). An even more intensive plate series, that of Lalande 21185 (BD + 36° 2147) was not included because a small, not yet analyzed perturbation is indicated both astrometrically (Lippincott and Hershey, 1982) and from infrared speckle interferometry (McCarthy and Hershey, 1982). Of the remaining 6 stars, 4 known visual binaries and two established perturbations; the orbital motions were allowed for and the residuals refer to the barycenter.

The residuals from the ten stellar paths were combined into weighted mean values for calender years over the interval 1937–1980. The forty-two yearly weighted mean residuals have total night weights ranging from 14 to 514. To arrive at a more nearly equal weight distribution, these means were grouped into fourteen normal points, ranging in total weight from 241 to 514, with an average value of 370, each covering one to four years. The results are shown in Figure 15.1 (solid dots). The resulting profile is remarkably linear ('horizontal'). The average deviations from zero for the fourteen points amount to

$$\left. \begin{array}{l} \pm\ 0.084\mu = \pm\ 0\rlap{.}''0016 \text{ in RA} \\ \pm\ 0.081\mu = \pm\ 0\rlap{.}''0015 \text{ in Decl.} \end{array} \right\} \quad \text{average } 0\rlap{.}''0016\,.$$

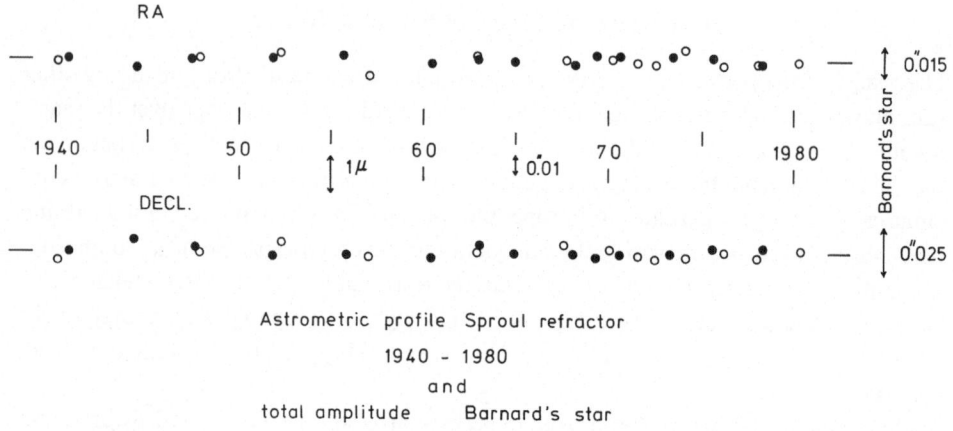

Astrometric profile Sproul refractor

1940 - 1980

and

total amplitude   Barnard's star

Fig. 15.1

The astrometric profile of the Sproul refractor therefore appears to have remained quite constant for these ten red stars over the interval 1937–1980. For these stars the average color of the reference stars is not red but rather yellow corresponding to two or three spectral classes earlier than M. The color correction over the interval 1942–1949 had been determined and applied; apparently there remains no color equation in the astrometric profile of the Sproul refractor.

A test was also made from less intensive series of thirteen non-red stars ranging in spectrum from A to G5, namely:

> Mu Cassiopeiae, Chi'Orionis, Zeta Herculis, 85 Pegasi, Mu Orionis, A 570, Ho 296, ADS 5234, ADS 8189, ADS 8939, ADS 9378, ADS 11060 and ADS 15176.

All are established visual binaries, except Mu Cassiopeiae. These stars were studied by Sarah Lee Lippincott (1975, 1982) and by John L. Hershey (1982). Observations over the interval 1937–1981 were made on a total of 1320 nights with a total night weight of 2930. The residuals from solutions for parallax, proper motion and orbital motion were grouped to obtain normal points of comparable weights. This meant combining successive years up to as many as seven for the earlier 'lean' years, partly due to competition on the observing program from 'single' stars. Thirteen normal points were formed ranging in total night weight from 155 to 275, with an average value of 225. The results are shown in Figure 15.1 (open circles).

As in the case of the ten red stars, the astrometric profile for the thirteen non-red stars in strikingly linear. The average deviations from zero for the normal points amount to:

$$\left. \begin{array}{l} \pm\,0.114\mu = \pm\,0\rlap{.}''0022 \text{ in RA} \\[2mm] \pm\,0.144\mu = \pm\,0\rlap{.}''0027 \text{ in Decl} \end{array} \right\} \quad \text{average } 0\rlap{.}''0025\,.$$

The twenty-three stellar paths dramatically demonstrate the constant long-range linear astrometric profile of the Sproul refractor. Twenty-seven normal points with an average deviation around 0.1 micron or $\pm$ 0.''002 attest to the virtual constancy of the astrometric portrayal. There is therefore good reason for satisfaction with the long-range astrometric stability of the Sproul refractor. The color equation over the interval 1941.82 to 1949.21 is now under control and can be allowed for. The correction for collimation in 1957 and the general renovation of the telescope in 1966 (Wanner, 1968) apparently have not affected the astrometric profile (Lippincott, 1978). The current profiles are in good agreement with other recent determinations (1981, 1982b; Lippincott, 1978; Hershey *et al.*, 1980).

Independant support for the long-range stability of the Sproul refractor is derived from a statistical study of periods and amplitudes of both resolved and unresolved astrometric binaries (Hershey *et al.*, 1980).

C. Perturbation of Barnard's Star

At Sproul Observatory Barnard's Star was first photographed over the interval 1916–1919, on 21 nights, to obtain a parallax determination. Beginning 1938 the star was observed on an average of thirty nights during each annual observing season. By 1980 the star had been photographed on 1200 nights; generally four plates each with up to five consecutive exposures were obtained on any one night. In 1916 the basic exposure time was 15 min, but is now less than 30 s; such has been the increase in sensitivity of photographic emulsions. Three background stars were adopted, all plates were measured on the Grant two-screw measuring machine (*Stellar Paths*, p. 10).

A recent analysis of the measurements was made for parallax, proper motion and secular acceleration. The resulting values are relative to the background of three reference stars whose proper motions have a marked effect on the secular acceleration. However a knowledge of the proper motions of the reference stars permits us to allow for this effect; how this is done beyond the scope of this treatise but may be found in *Stellar Paths*. A value of 0.''001 30 $\pm$ 0.''000 03 is found for the secular acceleration in sufficient agreement with the predicted value of 0.''001 25 $\pm$ 0.''000 02 (*Stellar Paths*, p. 54). The early observations (1916–1919) played a significant role in determining and strenthening the value of the acceleration effect.

Proper motion, acceleration and parallax were removed from the observed path (the measurements over the interval 1942–1948 were adjusted in the horizontal equatorial coordinate, right ascension, to allow for an instrumental color effect; for details see *Stellar Paths*, Chapters 3 and 18). The remainders, when combined into yearly averages ('normal points'), next were reduced to the center of mass of the solar system, thus eliminating any perturbing effects of the planets, particularly the most massive planets Jupiter and Saturn (Chapter 3).

The adjusted yearly remainders clearly reveal a pattern, the continuity of which points to consistent behavior of telescope and measuring apparatus (Figures 15.2 and 15.3). The policy of intensive observation on as many nights as possible over an intensive time

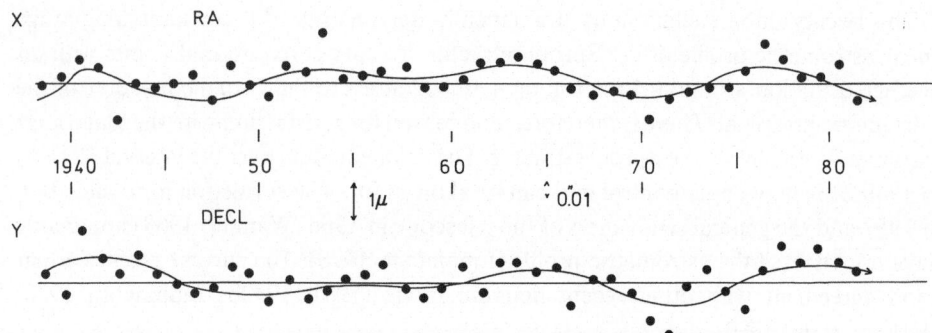

Barnard's Star 1938 - 1981 Sproul Observatory

Yearly normal points represented by

two circular orbits ; periods 12 and 20 yr.

Fig. 15.2

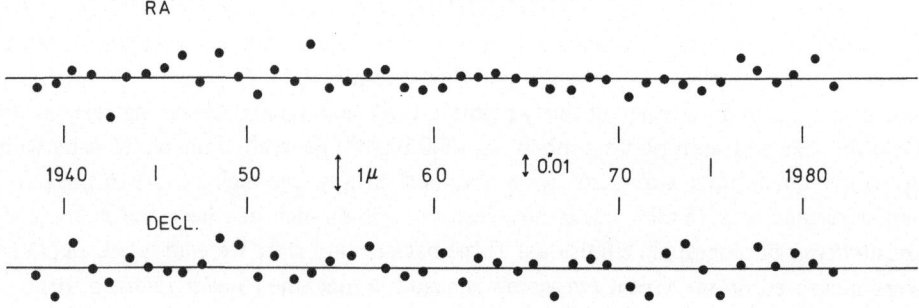

Barnard's star   1938 - 1981

Residuals from two orbits,

12 and 20 yr.

Fig. 15.3

interval, in this case more than four decades (1938–1981) is thus supported. The pattern of the remainders, can be best represented by a combination of a 'short' and a 'long' periodicity, each with a total amplitude somewhat below $0\rlap{.}''02$. The perturbation effect is too small to permit the recognition or determination of the shape (ellipticity) of the perturbations; within the accuracy of the material they are satisfactorily represented by *circular* orbits. The following results are adopted:

	'short' period	'long' period
Period	12 yr	20 yr
α	$0\rlap{.}''0070$	$0\rlap{.}''0064$
	0.013 AU	0.012 AU
Orbital constant $\alpha P^{-2/3}$	0.0025	0.0016

Note the very small values for the orbital constants. For an adopted value of 0.14 solar masses for Barnard's star, the masses of the companions are found to be (Chapter 11, Equation (11.18))

for the 'short' orbit 0.000 61 solar mass or 0.7 Jupiter mass;
for the 'long' period 0.000 43 solar mass or 0.5 Jupiter mass.

Both masses have probable errors of 0.1 Jupiter mass. Considering the minute perturbation effects, about one micron in the focal plane, there is some reason for guarded satisfaction. Early analyses (1963, 1969) resulted in considerable uncertainty in the interpretation; the values of the 'long' period ranged from 18.5 to 26 years, and the 'short' periodicity was not clearly recognized till 1969. By now the 44 consecutive years appear to permit a fair separation of the 'short' period covered more than three times, and the 'long' period covered more than twice. As a matter of fact successive analyses made since 1975 have essentially yielded the same results from which predicted future positions were close to those observed.

For the adopted value of 0.14 solar masses for the parent star, the short and long orbits of the two inferred companions around Barnard's star are found to have radii of 2.7 and 3.8 AU (Chapter 11, Equation (11.19)) and hence greatest separations from the star of 1″.5 and 2″.1, respectively. No trace of companions has been detected on the well over ten thousand exposures confirming the conclusion based on the very minute orbital constant (Chapter 11, Equation (11.17)) that the companions contribute no light to the measured images.

We conclude therefore, that the two planetary companions of Barnard's star have masses somewhat below that of Jupiter. Since Barnard's star has only 0.000 46 times the Sun's luminosity the apparent magnitudes due to reflected light would be about 30. They would therefore be too faint to be detected by direct visual observations with existing instrumentation. The next several years should witness substantial improvement for the values of the amplitudes and periods of the two perturbations, hence of the masses of the two companions. And, it remains to be seen, whether through drastic improvements in direct observing techniques, the companions might be 'seen'. Meanwhile the power of gravitational detection of unseen objects, appears to have revealed itself again.

D. Dynamical Interpretation, Planetary System of Barnard's Star

Figures 15.4 and 15.5 show the perturbation orbits and the orbits of the companions relative to Barnard's star. Adopting a mass of 0.14 $\mathcal{M}_\odot$ for Barnard's star, Kepler's third law yields orbital radii of 2.7 and 3.8 AU. A comparison with the values 0.013 and 0.012 AU for the radii of the perturbation orbits (Section 2) yields mass-ratios 1 to 210 and 1 to 320, relative to Barnard's star: the masses of the two unseen companions are 0.7 and 0.5 times Jupiter, respectively. Earlier determinations of the companion masses range from 0.7 $\mathcal{M}_2$ to 1.0 $\mathcal{M}_2$ for the 'short', 0.4 $\mathcal{M}_2$ to 0.6 $\mathcal{M}_2$ for the 'long' orbit (1975, 1977, 1979, 1980, 1981, 1982a).

Greatest separations of 1″.5 and 2″.1 from the primary are thus predicted; no trace

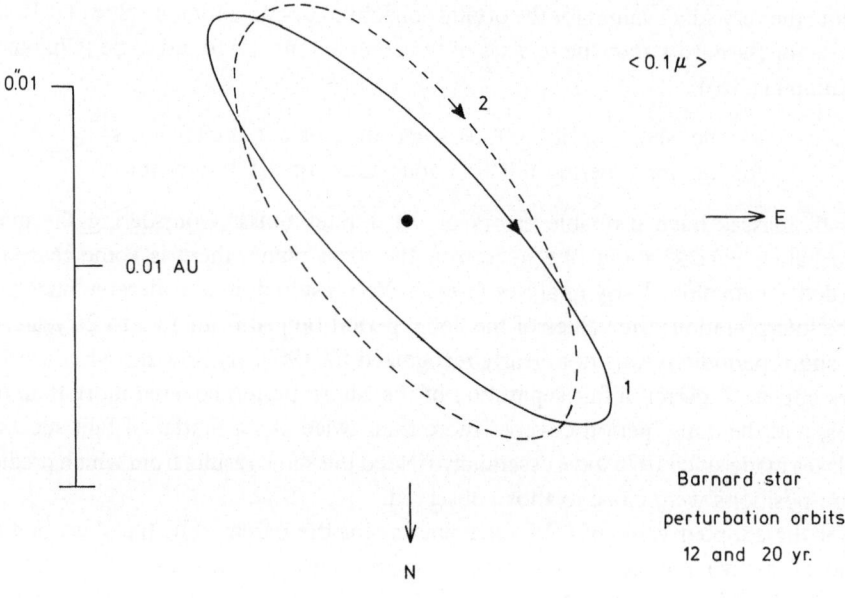

Fig. 15.4

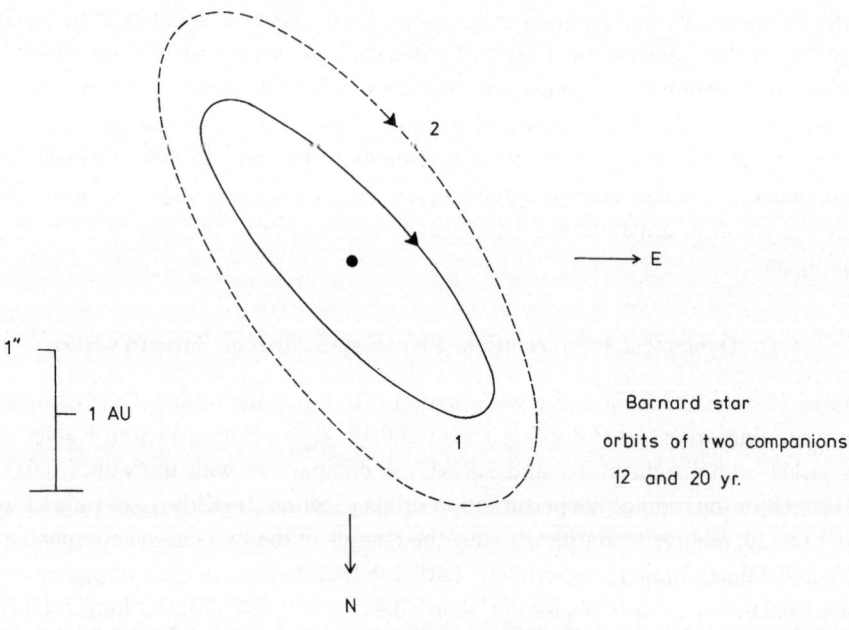

Fig. 15.5

of companions has been detected on the close to twenty thousand exposures. It is interesting in this connection to note the conclusion by McCarthy, Low and Kleinmann (Troy meeting, Americal Astronomical Society, June 1982) from an infrared study by means of speckle interferometry. While this study was successfull in detecting unseen companions for several stars, the results for Barnard's star were negative, indicating that at wavelength 2.2 microns any companions must be fainter than absolute magnitude 12.5. Since the corresponding absolute magnitude of Jupiter is about 24, these negative results do not exclude the existence of Jupiter-like companions as indicated by the Sproul astrometric material.

The relative inclination I of two orbits is given by

$$\cos I = \cos i \cos i_2 + \sin i, \sin i_2 \cos(\Omega_1 - \Omega_2).$$

While i_1 and i_2 are uniquely defined, the values of Ω_1, or Ω_2 may be changed by 180°, and two different values of I are found. One is 11°, earlier values range up to 34°, with the two companions co-revolving, the second is 138° ($-42°$), with the two companions counter-revolving; the probable errors of these inclinations are about $\pm 10°$.

Had we chosen not to recognize and allow for any perturbation in the path of Barnard's star, the average deviation (mean of RA 1nd Decl.) for the normal points of Barnard's star would be $\pm 0.36\mu = \pm 0''.0068$. From this we still could have concluded that the Sproul telescope had been essentially stable at the fractional micron level. But in view of the profile results from 23 stars, we preferred to accept a cosmic explanation for the systematic pattern of the yearly mean residuals for Barnard's star.

The observed perturbation of Barnard's star has a total amplitude of $0.82\mu = 0''.015$ in RA and $1.30\mu = 0''.025$ in Decl., some ten times larger than the average deviation of the normal points in the instrumental profile (Figure 1). The remarkable accuracy of the linear astrometric profile, frees the telescope from any suspicion that systematic changes in the performance of the instrument could be the cause of the inferred perturbation of Barnard's star, and of other stars; the orbital elements are 'pure' and of cosmic origin. The relevation of the perturbation of Barnard's star is a tribute to the instrument. To quote the late Joseph Ashbrook (letter of July 7, 1978) 'the telescope is the hero of the story'.

I see no observational approach at this time, which would permit a choice between the two alternatives. But I see no reason at the moment not to chose or prefer the first solution. To realize that the two planets may be *co-revolving* in near *co-planar*, *near-circular* orbits is not uninteresting. It could be the first intimation of an *extra-solar planetary system*, of which, by the only method available at this time, the first two, most massive 'Jovian' planets, have been found. The mossibility of circular orbits, and their coplanetary support the disk origin of planets (Chapter 8). This would be a situation analogous to our own solar planetary system, for which Jupiter and Saturn would be the first planets to be detected by the same astrometric technique from an observing station near Barnard's star. However the range in amplitudes would be three times smaller, due to the differing mass ratios, of primary and companion(s). Compare namely the aforementioned ranges in the solar path, $0''.008$ in RA, $0''.004$ in Decl. as observed from Barnard's star, with the range 0.″015 in RA, 0.″025 in Decl., observed by us for the perturbation of Barnard's star.

Speculating a little further. Could this be the first 'tip (or tips) of the iceberg'? How many planetary systems may be ready for discovery, and are patiently awaiting astrometric studies, by conventional photographic or other techniques? We shall always be confronted with treshold limitations, a natural condition of research, very wuch so of the present long-focus photographic astrometric approach. But should we not come to the rescue of a cosmic phenomenon which may be trying to reveal itself in a sea of errors?

Acknowledgement

For assistance and cooperation I am indebted to Sarah Lee Lippincott and John. L. Hershey, to Mary Jackson and other staff members of the Sproul Observatory, as well as to the Astronomical Institute 'Anton Pannekoek' of the University of Amsterdam. The continued support from the National Science Foundation is gratefully acknowledged.

CHAPTER 16

EPILOGUE

A. The Cosmic Experiment

We may regard the cosmic universe as a huge experiment in mechanics, physics and chemistry. The fact that this is running without our efforts and beyond our control, puts a perculiar responsibility on the astronomer and his approach. The observational material is based on radiation from the stars, which communicates messages at the crawling, yet unsurpassed speed of $300\,000$ km s^{-1}. Our standpoint of observation is essentially tied to planet Earth, and we are handicapped both by having to observe from the bottom of an ocean of air, the atmosphere, and by the gravitation at the surface of the Earth. Through research carried out from artificial satellites, we are overcoming some of these difficulties. On the other hand the Earth is an exeedingly stable observing platform, provided by nature and continues to play a major role in astronomical observations. The Milky Way system contains something like $50\,000$ million stars. Beyond, millions of similar systems are known to exist. Hundreds of thousands of stars in our Milky Way system are subject to individual scrutiny by the astronomer; each of these spheres of gases represents an experiment of tremendous magnitude.

The cosmic experiment consists, therefore, of myriads of stars and other celestial bodies spread through space and time, their motions governed by the universal law of gravitation and their radiation by the fundamental laws of atomic and nuclear physics. A beautiful exhibit is the mechanism of our solar system; the circuits of the planets about the Sun take place with a precision which in the long run may be far more accurate than that of any man-made clock.

Stars and nebulae provide opportunities for physical and chemical studies beyond the limits of the laboratory. Because of their high temperatures and the huge number of atoms, the interiors are fascinating largescale extensions of atomic experiments made on Earth. Nebulae and outer stellar atmospheres with their exceedingly low pressures exhibit physical conditions not duplicated on Earth.

Progress in scientific problems often takes place very slowly; the historical order of discovery is not necessarily the one which we would logically prefer in listing the results. The scientific approach introduces a systematic, though artificial dissection of the subject matter, which has no real significance as far as the structure of the material universe is concerned. It involves detailed planning and organization of separate investigations, each determined by the inherent limitations of method, instruments, the time and space allotted to man, his ability and imagination. We have intimations, obtain glimpses, carefully measure and check various revelations, which, in combination with other evidence, gradually give us some insight into a problem of overwhelming complexity.

It is the task of the astronomer to disentangle the various limitations and complications inherent in our very methods of observation and in our location. We make use of the methods of scientific inquiry, and observe, record our observations as accurately as possible. When the time is ripe, we attempt to interpret these observations and propose tentative models or hypotheses; these in turn lead to predictions which are tested by further observations. There must be a sound balance between theory and observation, while preconceived notions about what the universe should be like must be carefully avoided.

We cannot possibly observe all celestial objects all of the time. We therefore plan, and select, and thus limit our observational activities. The element of *chance* or *serendipity* must be recognized. Important discoveries are sometimes made *unplanned*, accidentaly, while looking for something else. The discovery of aberration (Chapter 3) is a classical example; another is the discovery of the unseen companion of VW Cephei (Chapter 14). An important quality of the astronomer remains therefore receptiveness toward the revelations, which the cosmos continually gives us a chance to recognize, by planning but always by us remaining alert for surprises.

The cosmic experiment points to the generally slow rate i.e. stellar lifetimes of the order of 10 000 million years. The astronomer is therefore aware of the need for conditions favorable to continued scientific endeavor, both in space and time. The limited span of life is overcome by a continued transfer of problems and knowledge from one generation to another. The one-world concept is beyond question; national boundaries are artificial. During the Second World War continued co-operation in astronomy was carried on as much as possible, without regard to the divisions between friend and foe.

The place of applied science is recognized, but the latter can grow only as the result of the desire for an absolute freedom of pure science.

Receptiviness toward what the Universe, or creation may reveal to us, remains an important quality of the scientist. Meanwhile the astronomer with 'two feet on Earth' will try to find out whether stars may have planets of some kind. There are numerous theories which attempt 'explain' the origin of the 'dependents', or the orbiting companions, planets *et al.*, of our star, the Sun. Why shouldn't other stars benefit from these theories? From a theoretical, or a speculative, point of view, there seems to be no valid reason why some or even all stars at some time may not have dependents in the form of planets. This is only a feeling and not a scientific attitude. 'Why' or 'why not' can only be settled observationally, not by theory or speculation which however may provide aid, stimulation and controls for the observing astronomers. Our Sun seems to be a single star, possibly an exceptional status in a universe where duplicity and multiplicity among stars are prevalent.

As far as we know, none of the other solar planets is a suitable abode for life as we know, is on Earth. And even here things are marginal; if the Earth as a whole were a few degrees warmer or colder, there would be severe changes in climatic and living conditions. Are 'we' the only astronomers in the solar system? And what intelligent life beyond; are there planets out there? One of the purposes of this treatise is to look into the possibility of planets near other stars; (Part 2). Whatever knowledge has been acquired so far, we have not indications yet of life – intelligent, or otherwise outside our solar system. One may speculate about that, of course – one even should –, one should not stifle on's imagination. Possibly 'space probes' and interplanetary communication may eventually help out, both trying to attain 'them' or 'we' being receptive in case 'they' might be trying to reach us.

B. Space and Time

The *basic observations* in cosmic kinematics are the spatial components observed at certain *times*. The *spatial* components are *geometric* and require appropriate observations of distance and directions relative to the Sun or the barycenter of the solar system. These astrometric data are obtained by appropriate instruments of which in the present treatise and specific problems, long-focus telescopes are basic examples. For nearby orbital paths such as moon and artificial satellites the time measures require very high accuracy. The same is true for stellar orbits with extremely short cycles down to periods of less than one year such as for bineries.

In *astrometric* studies of visual binaries (whether observed visually, photographically, or by other advanced techniques), the time element may be regarded in a more relaxed manner. For binaries with long periods, say of the order of decades or centuries, the time element, to match the accuracy of the geometric information, need not be known with an accuracy better than a week or so. Fot shorter periods, of the order of one year, and this obviously includes parallax measurements performed from the Eerth's orbit around the Sun, a higher accuracy is required though it need not be better than say one hour i.e. about $0''.0001$/year; this corresponds to $0''.001$ in the annual proper motion of Barnard's star (Chapter 15). For the specific problem of perturbation in astrometrically measured stellar paths time measurement need even be less accurate, though one may want to maintain the above mentioned limit of $0''.0001$/year, since parallax may be involved. For the problems we are concerned with in this treatise 'any' clock therefore will do, since the cycles of the phenomena are rarely below one year and also the geometric accuracy is limited. This cannot be stressed enough since a large time span (~ 10 to 100 years or more), may be an inherent property of our particular problem. Therefore as far as timing is required, long-time intervals may be of the essence. An accuracy of one day generally is sufficient, but an extension of any time span generally becomes desirable. In other words continued observations become a most important ingredient, and an ultimate desideratum. The time component requires only the simplest measurement, obtained at no expense or effort, but readiness, inexorably is required, to extend the observations over many human generations. And while the geometric components may become more accurate, with the application of new techniques, the time element cannot and need not be improved.

We repeat, of the space-time elements the most accurate and most easily obtained is time. Time will always work for us provided we realize this very obvious fact and live accordingly: 'hâtez-vous lentement!' Therefore: *patience* and *perseverence* are of the essence; let time work for us, by our waiting for the effect of the cheapest parameter, time which *cannot* be speeded up by any fancy human technology. In all these studies of stellar paths do not forget the importance of the time element. Whatever technological advances are made in the possible detection of unseen companions, no dynamical information and cosmogonic explanations, can be made till *orbital* elements are obtained. And for this purpose, time is of the essence. I am thinking particularly of phenomena, where periods of decades or more, even centuries are involved. The time element cannot be mastered except through patience. Apart from the necessity of waiting, the time-element is by for the most accurate orbital parameter, furnished by nature as a reward for our patience.

C. Astronomy and Astronomers

One might say: Astronomy is 'what astronomers do' and these doings change from century to century, if not from decade to decade, or even from year to year. After having been engaged in research and teaching for over more than half a century, I have been struck and impressed by the, at times drastic, changes in global astronomical activities.

Why is this so? The Universe has remained the same! Are there changing fashions in astronomy? Yes. Part of this is a natural scientific development, and there is nothing wrong with lifting out of the mass of available material objects of particular interest. Both the professional and laymen's interest in white dwarfs were followed by a shift to neutron stars, pulsars while black holes were on front stage later. The expanding universe was the rage fifty years ago; as a matter of fact, it still is. Origin and avolution, new techniques, radio, X-ray. All very exciting, and how to retain a sound overview. While basic tools of the astronomer remain mathematics, mechanics, physics and chemistry, more than that is required. *Imagination, persistence, curiosity, patience,* and *faith est sine qua non.*

I should like to stress the significance of the relatively small number of astronomers, who as outstanding figures, have played leading roles in 'passing on the torch of learning' over the ages. With due respect to technology and teamwork where it is needed, I cannot fail to be impressed by the traditional handful of the few special people who have been giants and milestones in the development and progress of astronomy. These individuals accomplished breakthroughs through innate education, imagination, vision and hard work, also by living at the right time and having a bit of luck. And they had a great influence on an overwhelming majority of others who often without realizing it (or willing to admit it), were waiting for the influence of these breakthroughs, and opportunities thus created. Restricting myself to some recent names such as Kapteyn, Russell, Eddington and Shapley. I believe that in the absence of these individuals, their breakthroughs sooner or later, would have been accomplished by others; such appears to be inevitable course of events in science. The merits of the great astronomers are that they had the required qualities, plus compulsion and vision which lifted them above the mass of people active in the field. When needed, they had courage in the face of expected opposition based on the, naturally conservative establishment, which at any one time slows down the acceptance of something new. It is exactly the new, sometimes, revolutionary results and ideas, which are life blood of continuing development of the field of study and research.

How different are developments in other fields, such as literature and the arts!

Let us be grateful to fate, for providing at any one epoch humans of elite quality, whether in science, music, or any other field, people who have the gift and the message to contribute to what makes life worth living.